T0155591

NEUROMETHODS

Series Editor
Wolfgang Walz
University of Saskatchewan
Saskatoon, SK, Canada

For further volumes:
http://www.springer.com/series/7657

Neuromethods publishes cutting-edge methods and protocols in all areas of neuroscience as well as translational neurological and mental research. Each volume in the series offers tested laboratory protocols, step-by-step methods for reproducible lab experiments and addresses methodological controversies and pitfalls in order to aid neuroscientists in experimentation. *Neuromethods* focuses on traditional and emerging topics with wide-ranging implications to brain function, such as electrophysiology, neuroimaging, behavioral analysis, genomics, neurodegeneration, translational research and clinical trials. *Neuromethods* provides investigators and trainees with highly useful compendiums of key strategies and approaches for successful research in animal and human brain function including translational "bench to bedside" approaches to mental and neurological diseases.

Imaging and Quantifying Neuronal Autophagy

Edited by

Ben Loos

Department of Physiological Sciences, Stellenbosch University, Stellenbosch, South Africa

Esther Wong

Healthy Longevity Translational Research Programme, Department of Physiology, Yong Loo Lin School of Medicine, National University of Singapore, Centre for Healthy Longevity, National University Health System, Singapore, Singapore

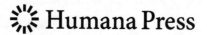

Editors
Ben Loos
Department of Physiological Sciences
Stellenbosch University
Stellenbosch, South Africa

Esther Wong
Healthy Longevity Translational Research Programme
Department of Physiology, Yong Loo Lin School of Medicine
National University of Singapore, Centre for Healthy Longevity
National University Health System
Singapore, Singapore

ISSN 0893-2336   ISSN 1940-6045   (electronic)
Neuromethods
ISBN 978-1-0716-1591-1   ISBN 978-1-0716-1589-8   (eBook)
https://doi.org/10.1007/978-1-0716-1589-8

This Humana imprint is published by the registered company Springer Science+Business Media, LLC part of Springer Nature.
The registered company address is: 1 New York Plaza, New York, NY 10004, U.S.A.

Preface to the Series

Experimental life sciences have two basic foundations: concepts and tools. The *Neuromethods* series focuses on the tools and techniques unique to the investigation of the nervous system and excitable cells. It will not, however, shortchange the concept side of things as care has been taken to integrate these tools within the context of the concepts and questions under investigation. In this way, the series is unique in that it not only collects protocols but also includes theoretical background information and critiques which led to the methods and their development. Thus it gives the reader a better understanding of the origin of the techniques and their potential future development. The *Neuromethods* publishing program strikes a balance between recent and exciting developments like those concerning new animal models of disease, imaging, in vivo methods, and more established techniques, including, for example, immunocytochemistry and electrophysiological technologies. New trainees in neurosciences still need a sound footing in these older methods in order to apply a critical approach to their results.

Under the guidance of its founders, Alan Boulton and Glen Baker, the *Neuromethods* series has been a success since its first volume published through Humana Press in 1985. The series continues to flourish through many changes over the years. It is now published under the umbrella of Springer Protocols. While methods involving brain research have changed a lot since the series started, the publishing environment and technology have changed even more radically. Neuromethods has the distinct layout and style of the Springer Protocols program, designed specifically for readability and ease of reference in a laboratory setting.

The careful application of methods is potentially the most important step in the process of scientific inquiry. In the past, new methodologies led the way in developing new disciplines in the biological and medical sciences. For example, Physiology emerged out of Anatomy in the nineteenth century by harnessing new methods based on the newly discovered phenomenon of electricity. Nowadays, the relationships between disciplines and methods are more complex. Methods are now widely shared between disciplines and research areas. New developments in electronic publishing make it possible for scientists that encounter new methods to quickly find sources of information electronically. The design of individual volumes and chapters in this series takes this new access technology into account. Springer Protocols makes it possible to download single protocols separately. In addition, Springer makes its print-on-demand technology available globally. A print copy can therefore be acquired quickly and for a competitive price anywhere in the world.

Saskatoon, SK, Canada   *Wolfgang Walz*

Preface

Autophagy flux and kinetics, machinery, cargo, rate and resolving power—these are the key terms that may characterize this book best. Macroautophagy (referred to here as autophagy) is an intracellular degradation process that involves the formation of a double-membrane vesicle, called autophagosome. Upon the sequestration of cytoplasmic cargo, autophagosomes are transported, dynein-dependently along the tubulin network, and subsequently fuse with lysosomes, where degradation of the cargo takes place. Mammalian cells undergo autophagy at a basal level, so as to preserve and maintain cellular homeostasis. Neurons rely on a particularly efficient autophagy machinery, and it is now clear that autophagy dysfunction plays a key role in the molecular pathology that characterizes certain neurodegenerative disorders. Although our understanding of the molecular machinery and associated signaling that is involved in autophagy regulation has grown tremendously, a major focus area that receives continuous attention centers around the importance of accurately assessing autophagy activity, particularly in the context of complex pathway dynamics, the kinetics of molecular machinery and cargo, the subcellular organization, and the discernment between activity and pathway intermediates. Microscopy techniques have contributed substantially to the discovery of the autophagy pathway and are contributing to unravel novel insights in dynamics, behavior, localization, and resolution. In light of ongoing advancements in imaging technologies, including fluorescence photo-activation, super-resolution and correlated light and electron microscopy, the autophagy pathway, its activity and molecular makeup, is being captured as never before. This volume is aimed to sequentially share the critical role of particularly fluorescence and electron microscopy techniques, to capture and quantify neuronal autophagy.

By bringing together a large number of experts, clinicians, microscopists, and molecular scientists, it is our hope to particularly unmask the dynamic nature of the autophagy pathway and suitable techniques that allow to capture and most sensitively quantify this process in neurons. It is also our hope that a strong sense for both machinery and its activity, but also cargo and its clearance rate, is being established, so as to enhance in vivo autophagy assessment and clinical translation. The volume therefore begins with fundamental, historical, and foundational approaches that began in baker's yeast, *Saccharomyces cerevisiae*, to guide and highlight the role of electron microscopy but also live cell imaging using fluorescently tagged autophagy proteins, which enabled to establish the order of their recruitment to the PAS, and, importantly, the rate of puncta appearance and its correlation with the rate of autophagosome formation. One may ask, why begin with a non-neuronal cell model? It is because the understanding of precise autophagosome identification, the measure of its formation rate, and its size control become clear in an unparalleled fashion and underpin many of the approaches that are implemented in the neuronal system. Wild-type yeast cells produce approximately 10–14 autophagosomes per cell per hour upon the first hours of nitrogen starvation. Neuronal autophagy flux is substantial in its magnitude, with a autophagosome generated every 3 min in primary neurons, but identification of autophagic structures is not trivial. In fact, literature reveals that basic problems remain in the correct identification of autophagic structures in transmission electron microscopy samples. Hence, a guide is provided for the identification of autophagic structures and their interpretation, applicable for the neuronal context, with reference to the appropriate sample preparation. Neuronal autophagy activity is spatially distinct, requiring spatiotemporal but also intracellularly regional resolution that includes somal, axonal, and dendritic compartments. These compartments can differ substantially in their autophagy complement, including differential regulation. Unique guidance is provided on the imaging of organelle dynamics in dorsal root ganglion (DRG), hippocampal, and cortical neurons isolated from either rat or mouse. Moreover, to rapidly discern neuronal autophagy flux has remained a challenge, and to do so without experimental disruption has remained equally nontrivial. Hence, methods are provided involving a recently developed autophagic flux probe, enabling rapid visualization and quantification of autophagic flux in neurons and zebrafish spinal cord.

Single-cell autophagy flux analysis approaches are provided by using optical pulse labeling (OPL), performed in primary rodent neurons. By using LC3-EOS, the protein half-live can be analyzed with high precision, over hours or even days. More so, a detailed method is provided to acquire image data allowing to quantify autophagosome flux, pool size, and the time required by the cell to clear its autophagosome pool. However, it has become increasingly clear that proteinaceous cargo is characterized by its characteristic receptor recruitment and subsequent clearance. This is of importance, so as to tune autophagy activity with the desired clearance of proteins associated with proteotoxicity and neurodegeneration. Hence, a detailed method is provided for the photoconversion and analysis of clearance kinetics of Dendra-tagged tau proteins, using neurons from larval zebrafish. Moreover, kymograph-driven methods are presented to assess and quantify "motor-sharing" mechanisms that drive autophagosome transport towards the soma. Next, methods to image and quantify neuronal autophagy using tdEOS-LC3 and the contribution to neuronal and dendritic morphogenesis, including spine density, using primary rodent hippocampal and cortical neuron cultures in the context of amyotrophic lateral sclerosis (ALS) and frontotemporal dementia (FTD) are described. Here, using GFP as a cell filler, the neuronal morphology and dendritic spines can be visualized and quantified in the context of autophagy. The final chapter aims to "close the circle" by highlighting the power of correlative light and electron microscopy (CLEM), enabling a coming together of the high resolution and, importantly, ultrastructural detail together with the precise information on specific macromolecules or proteins of interest, identifying neuronal autophagosomes and autolysosomes in both 2- and 3-dimensional neuronal context.

The chapters most often provide details on open image analysis platforms, such as Image J/FIJI allowing the processing and implementation of similar approaches without the requirement of specialized analysis software.

It is our hope that this volume will provide insights into the power of microscopy tools to image and quantify neuronal autophagy with high precision, encourage implementation of live cell imaging, photoactivation, and correlative techniques, and draw attention to the kinetics of both neuronal autophagy machinery and cargo in a practical and applicable manner.

This book is dedicated to the late Prof Beth Levine, who has been an inspirational colleague of our times.

Stellenbosch, South Africa   *Ben Loos*
Singapore, Singapore   *Esther Wong*

Contents

Contributors

STEVEN K. BACKUES • *Department of Chemistry, Eastern Michigan University, Ypsilanti, MI, USA*

NICHOLAS A. CASTELLO • *Center for Systems and Therapeutics & Taube/Koret Center for Neurodegenerative Disease, Gladstone Institutes, University of California San Francisco, San Francisco, CA, USA; Departments of Neurology and Physiology, University of California San Francisco, San Francisco, CA, USA*

KELLY A. CHAMBERLAIN • *Synaptic Function Section, The Porter Neuroscience Research Center, National Institute of Neurological Disorders and Stroke, National Institutes of Health, Bethesda, MD, USA*

XIU-TANG CHENG • *Synaptic Function Section, The Porter Neuroscience Research Center, National Institute of Neurological Disorders and Stroke, National Institutes of Health, Bethesda, MD, USA*

LUCY COLLINSON • *Electron Microscopy Science Technology Platform, The Francis Crick Institute, London, UK*

ANDRÉ DU TOIT • *Department of Physiological Sciences, Stellenbosch University, Stellenbosch, South Africa*

TOMOYA EGUCHI • *Department of Biochemistry and Molecular Biology, Graduate School of Medicine, The University of Tokyo, Tokyo, Japan*

LIZE ENGELBRECHT • *Central Analytical Facilities, Fluorescence Microscopy Unit, University of Stellenbosch, Stellenbosch, South Africa*

EEVA-LIISA ESKELINEN • *Institute of Biomedicine, University of Turku, Turku, Finland; Molecular and Integrative Biosciences Research Programme, University of Helsinki, Helsinki, Finland*

STEVEN FINKBEINER • *Center for Systems and Therapeutics & Taube/Koret Center for Neurodegenerative Disease, Gladstone Institutes, University of California San Francisco, San Francisco, CA, USA; Departments of Neurology and Physiology, University of California San Francisco, San Francisco, CA, USA*

ANGELEEN FLEMING • *Department of Medical Genetics, University of Cambridge, Cambridge Institute for Medical Research, Cambridge, UK; Department of Physiology, Development and Neuroscience, University of Cambridge, Cambridge, UK*

WAN YUN HO • *Department of Physiology, National University of Singapore, Singapore, Singapore*

JAN-HENDRIK S. HOFMEYR • *Department of Biochemistry, Stellenbosch University, Stellenbosch, South Africa*

ERIKA L. F. HOLZBAUR • *Department of Physiology, Perelman School of Medicine, University of Pennsylvania, Philadelphia, PA, USA*

MARTIN L. JONES • *Electron Microscopy Science Technology Platform, The Francis Crick Institute, London, UK*

LYDIA-MARIE JOUBERT • *Central Analytical Facilities, Electron Microscopy Unit, University of Stellenbosch, Stellenbosch, South Africa*

KATRI KALLIO • *Molecular and Integrative Biosciences Research Programme, University of Helsinki, Helsinki, Finland*

DANIEL J. KLIONSKY • *Department of Molecular, Cellular and Developmental Biology, Life Sciences Institute, University of Michigan, Ann Arbor, MI, USA*

JURGEN KRIEL • *Department of Physiological Sciences, University of Stellenbosch, Stellenbosch, South Africa*

SHUO-CHIEN LING • *Department of Physiology, Yong Loo Lin School of Medicine, National University of Singapore, Singapore, Singapore; Neurobiology/Ageing Programme, National University of Singapore, Singapore, Singapore; Program in Neuroscience and Behavior Disorders, Duke-NUS Medical School, Singapore, Singapore*

BEN LOOS • *Department of Physiological Sciences, Stellenbosch University, Stellenbosch, South Africa*

ANA LOPEZ • *Department of Medical Genetics, University of Cambridge, Cambridge Institute for Medical Research, Cambridge, UK; Department of Physiology, Development and Neuroscience, University of Cambridge, Cambridge, UK*

DUMISILE LUMKWANA • *Central Analytical Facilities, Fluorescence Microscopy Unit, University of Stellenbosch, Stellenbosch, South Africa*

NOBORU MIZUSHIMA • *Department of Biochemistry and Molecular Biology, Graduate School of Medicine, The University of Tokyo, Tokyo, Japan*

HIDEAKI MORISHITA • *Department of Biochemistry and Molecular Biology, Graduate School of Medicine, The University of Tokyo, Tokyo, Japan; Department of Physiology, Graduate School of Medicine, Juntendo University, Tokyo, Japan*

CHRISTOPHER J. PEDDIE • *Electron Microscopy Science Technology Platform, The Francis Crick Institute, London, UK*

DAVID C. RUBINSZTEIN • *Department of Medical Genetics, University of Cambridge, Cambridge Institute for Medical Research, Cambridge, UK; UK Dementia Research Institute, University of Cambridge, Cambridge Institute for Medical Research, Cambridge, UK*

ZU-HANG SHENG • *Synaptic Function Section, The Porter Neuroscience Research Center, National Institute of Neurological Disorders and Stroke, National Institutes of Health, Bethesda, MD, USA*

ANDREA K. H. STAVOE • *Department of Physiology, Perelman School of Medicine, University of Pennsylvania, Philadelphia, PA, USA; Department of Neurobiology and Anatomy, The University of Texas Health Science Center at Houston McGovern Medical School, Houston, TX, USA*

Chapter 1

Quantification of Autophagosome Size and Formation Rate by Electron and Fluorescence Microscopy in Baker's Yeast

Steven K. Backues and Daniel J. Klionsky

Abstract

The use of both transmission electron microscopy and fluorescence microscopy have provided tremendous advances to our understanding of autophagosome formation in baker's yeast, *Saccharomyces cerevisiae*. In the last decade, parallel techniques have been developed for both types of microscopy that allow the quantification of the rate of autophagosome formation. Importantly, these techniques, unlike other measures of total autophagic flux, allow a researcher to distinguish between effects on autophagosome size and autophagosome number. This has led to the discovery that certain autophagy proteins (e.g., Atg8) contribute primarily to the control of autophagosome size, whereas others (e.g., Atg9) are principally involved in controlling autophagosome number, suggesting different roles for these proteins in the autophagosome formation process.

In this chapter, we present two methods for quantifying autophagosome formation in yeast. One, based on electron microscopy analysis of autophagic bodies in the vacuole, can give estimates of both autophagosome size and number. The other, based on live-cell imaging of growing autophagosomes labeled with GFP-Atg8, can provide information on the rate of autophagosome formation. Together they provide a robust toolbox for analyzing the roles of different proteins in the process of autophagosome formation.

Keywords Autophagic body, Autophagosome, Autophagy, Fluorescence microscopy, GFP-Atg8, Transmission electron microscopy, Vacuole

1 Introduction: Detection of Autophagy by Microscopy in Baker's Yeast

Although the very first images of autophagy come from other organisms, imaging of autophagic structures in baker's yeast (*Saccharomyces cerevisiae*) has played a critical role in autophagy research past and present, supporting genetic and biochemical studies in this model organism [1]. It is for this reason that this chapter shall serve as a critical introduction to both electron microscopy and fluorescence microscopy techniques that contribute to the various approaches implemented in imaging and quantifying neuronal autophagy. These yeast studies began with transmission electron microscopy (TEM), much of it performed by the talented microscopist Dr. Misuzu Baba, starting with the original papers that

Ben Loos and Esther Wong (eds.), *Imaging and Quantifying Neuronal Autophagy*, Neuromethods, vol. 171, https://doi.org/10.1007/978-1-0716-1589-8_1, © Springer Science+Business Media, LLC, part of Springer Nature 2022

established the occurrence of starvation-induced bulk autophagy in yeast [2], and showed that the ultrastructure of this process is "essentially similar" to that of autophagy in mammals [3]. These studies showed that autophagy in yeast begins with a "cup-shaped" membrane structure, the phagophore, which envelops cytoplasm and then closes into a complete double-membraned autophagosome. The outer membrane of the autophagosome subsequently fuses with the limiting membrane of the vacuole, the yeast equivalent of the lysosome, thereby releasing the inner vesicle into the vacuole lumen; this single-membrane structure is referred to as an autophagic body [3]. The membrane of the autophagic body is lysed, allowing the contents of the autophagic body to be degraded and recycled [3]. Using similar methods, researchers showed shortly thereafter that the precursor form of the vacuolar protease Ape1 (prApe1) follows a similar pathway to the vacuole [4]. These studies established the first example in this organism of selective autophagy, which uses the autophagic machinery to transport specific cytoplasmic cargos into the vacuole/lysosome. The prApe1 is transported as a large oligomeric complex that is electron dense and easily visible by TEM [4], and can thus provide a key landmark for detecting forming autophagosomes in TEM images in yeast [5, 6].

Yeast TEM is not only of historical importance but also continues to deliver key insights into the process of autophagosome formation. For example, recent ultrastructural analysis of certain autophagy mutants by Dr. Baba has led to the identification of the "alphasome," which may be a very early autophagy intermediate derived from the nuclear membrane [6]. Similarly, recent EM analysis of Atg2 mutants from the lab of Dr. Fulvio Reggiori has helped to establish a role for Atg2 in connecting the autophagosome to the ER and supporting phagophore expansion [7].

One of the ongoing uses of TEM in yeast is to shed light on the process of autophagosome formation by determining the size and number of autophagosomes formed in different yeast mutants or under different environmental conditions. Various techniques can be used to measure total autophagic flux [8–10], but only electron microscopy has the resolution necessary to measure the size of individual autophagosomes, and thus to distinguish between changes in size and changes in number [11]. Autophagosomes typically fuse with the vacuole almost as soon as they are formed. Consequently, completed autophagosomes in the cytoplasm are rare and difficult to capture even under strongly inducing nitrogen-starvation conditions [3]. In contrast, inhibition of vacuolar protease activity allows the autophagic bodies and their contents from multiple autophagosomes to accumulate inside the vacuole after fusion. Therefore, autophagic bodies have served as a proxy for autophagosomes ever since the very first description of yeast autophagy [2]. In addition, depending on the fixation methods it is possible for the vacuole to appear as an essentially blank or white compartment, making it easier to distinguish and evaluate autophagic bodies compared to autophagosomes that are present in the relatively dense cytosol.

Analysis of autophagic bodies formed in yeast with alterations in the levels or activity of specific autophagic proteins has demonstrated that autophagosome size (but not number) is particularly controlled by the amount of functional Atg8 in the cell [11, 12], whereas autophagosome number (but not size) is controlled primarily by the amount of Atg9 delivered to the site of autophagosome assembly [13–17], suggesting critical but distinct roles for these two proteins in phagophore initiation and expansion. Interestingly, Atg7, an upstream protein in the Atg8 pathway, affects both autophagosome size and autophagosome number, suggesting that it may have additional Atg8-independent roles [18]. Methods to accurately estimate the size and number of autophagic bodies from TEM sections have been developed by workers in the Klionsky lab [19, 20] and are presented in this chapter.

Another type of microscopy that has given critical insight into autophagy in yeast, as with other systems, is fluorescence microscopy. In particular, fluorescent tagging of known autophagy proteins allowed the identification of the phagophore assembly site (PAS) [21, 22], the site in the yeast cell adjacent to the vacuole where autophagosomes are formed. Although some autophagy proteins such as Atg9 and its machinery show steady-state localizations to other subcellular compartments [14, 23–25], at least a subpopulation of every core autophagy protein—those required for autophagosome formation—will localize to the PAS at some point during its lifecycle. Moreover, live-cell imaging using fluorescently tagged autophagy proteins has helped to establish the order in which they are recruited to the PAS, thus shedding light on the overall pathway of autophagosome formation [26].

Given the centrality of the PAS in yeast autophagy, it is critical to have well-defined fluorescent markers of this structure. The two most commonly used PAS markers are fluorescently tagged versions of prApe1 and Atg8. As a selective autophagy cargo, prApe1 is useful because it forms a stable complex, and thus RFP-Ape1 forms a single bright punctum in the cell that is found at the site of autophagosome formation under starvation-inducing conditions [27]. However, RFP-Ape1 would not be expected to mark every autophagosome under strongly inducing conditions. In contrast, Atg8, a small ubiquitin-like protein that is covalently attached to phosphatidylethanolamine in the phagophore membrane, is thought to be involved in every canonical autophagosome formation event. This is why GFP-Atg8 has emerged as one of the most common PAS markers (*see* **Note 1**).

GFP-Atg8 puncta are rare in nutrient-rich conditions where autophagy is repressed, but during starvation conditions most cells contain a visible GFP-Atg8 punctum, and sometimes even more than one. These puncta represent autophagosomes forming at the PAS, and in most cases the disappearance of the punctum reflects release of GFP-Atg8 from the autophagosome during a late step in its formation [11]. Live-cell imaging of GFP-Atg8 puncta can be used to measure the time necessary for formation of a single autophagosome, and the rate of puncta appearance should correlate with the rate of autophagosome formation [13, 19]. This then presents an independent measure of autophagosome number that can be correlated with data acquired from TEM imaging; methods for measuring the rate of autophagosome formation by fluorescence microscopy with GFP-Atg8 are also presented in this chapter.

2   Materials

2.1   Embedding and TEM

To prevent breakdown of autophagic bodies in the vacuole, we used protease-deficient *pep4Δ* yeast strains; these strains were also *vps4Δ* to prevent background signal within the vacuole from small vesicles from the multivesicular-body pathway [28]. $KMnO_4$ for fixation (RT20200), dry (glass-distilled) acetone for dehydration (RT10016), and a Low Viscosity Embedding Media Spurr's Kit (14300) were purchased from Electron Microscopy Sciences (EMS). A nutating mixer (TSC Scientific Corp. Model 117) was used for sample incubations during all processing steps, as this is gentler than end-over-end rotation. We generated 70-nm ultrathin sections on a Leica UC7 ultramicrotome, collected them with a Perfect Loop (EMS, 70944) onto 300-mesh copper grids (EMS, EMS300-Cu), and stained them for 5 min with "Uranyless" (EMS, 22409) and then for 5 min with Reynold's Lead Citrate (EMS 22410), both ready-to-use in pump bottles. Samples were then imaged on a JEOL JEM 1400 TEM at the University of Michigan Microscopy and Image Analysis Laboratory. Outlines of vacuoles and autophagic bodies were drawn manually in Adobe Photoshop, measured with ImageJ (FIJI distribution) and the data analyzed with Microsoft Excel and R.

2.2   Immobilization and Imaging of GFP-Atg8 Cells Using Concavity Slides

In order to follow multiple rounds of autophagosome formation in real time, cells must be immobilized on a solid support and given sufficient media. We used concanavalin A-treated coverslips and concavity slides for this purpose. Single-depression concavity slides (1.5–1.95-mm deep) were purchased from Electron Microscopy Sciences (71878-07). Concanavalin A-treated coverslips were prepared by pipetting 100 µl of a freshly prepared 1% w/v solution of concanavalin A (Sigma, L7647) in water onto one side of a 24 × 60-mm rectangular coverslip (Fisher Scientific, S175201), spreading to cover the surface, and allowing the coverslip to stand at room temperature for 5 min. The coverslip was then rinsed five times by dipping in water and allowed to air dry. Cover slips were used promptly (no more than a few hours) after concanavalin A treatment.

Cells were imaged on a DeltaVision Elite deconvolution microscope (GE Healthcare/Applied Precision) with a $100\times$ objective and a CCD camera (CoolSnap HQ; Photometrics).

3   Methods

3.1   Acquisition of TEM Images

A detailed protocol for measuring the size and number of autophagic bodies by TEM is available elsewhere [20], so here we will outline our typical procedure and emphasize key aspects. Vacuolar protease-deficient yeast were cultured under standard conditions in rich medium (see Note 2). Non-selective autophagy was induced by 2–4 h of nitrogen starvation in SD-N medium (0.17% yeast nitrogen base without amino acids or ammonium sulfate [ForMedium, CYN0501], 2% glucose). Starved cultures (~30 OD_{600} units of cells) were harvested by centrifugation at $1600 \times g$, for 5 min, resuspended in 30 ml of purified water to wash, and harvested again at $1600 \times g$, for 5 min. To fix the cells, the washed pellet was resuspended in 1.5 ml of freshly prepared 1.5% $KMnO_4$ and transferred to a 1.5-ml microcentrifuge tube. The samples were incubated at 4 °C with nutation, first for 30 min and then with fresh 1.5% $KMnO_4$ overnight. Fixed cells were washed $4\times$ with water at room temperature then dehydrated with increasing concentrations of acetone (30–100% in six steps with 20 min nutation per step, then twice more in 100% acetone). Cells were embedded in increasing concentrations of Spurr's resin (1 h nutation in 20% and then 50% resin, then 8 h each in two changes of 100% resin), and polymerized overnight at 70 °C after pelleting in a 0.65-ml microcentrifuge tube. This yielded a conical block with a pellet of embedded cells at the tip that could be faced, trimmed, and sectioned into 70-nm ultrathin sections. These sections were then stained and imaged by TEM. Images were acquired at $6000\times$ for measurement of vacuoles and at $30,000\times$ for measurement of autophagic bodies within the vacuole. Images were captured systematically to avoid sampling bias; for example, every cell with a visible vacuole within a given grid square would be imaged (see Note 3). Autophagic bodies could be recognized as dark, cytoplasmic-containing structures surrounded by a dark membrane within the white vacuole (Fig. 1). At least 100 images were acquired per sample, with two independent samples per experimental condition (see Note 4).

3.2   Quantification of EM Micrographs

The first step in the analysis of the TEM images is to outline the vacuole and autophagic body cross-sections visible in the images so that they can be analyzed. Unfortunately, existing machine vision tools are not sufficient to reliably recognize the boundaries of these structures, so the outlining must be done manually. We did the

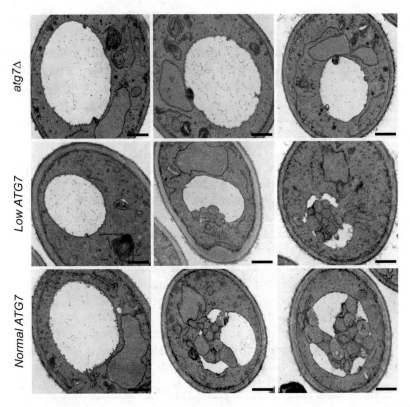

Fig. 1 Representative TEM micrographs of autophagic body sections. Strains expressing no Atg7 (*atg7Δ*), low levels of Atg7, and normal levels of Atg7 were starved for 3 h in SD-N, fixed with KMnO$_4$ and embedded in Spurr's resin. Ultrathin sections (70 nm) were acquired at 30,000× magnification by TEM. Few to no autophagic body sections are seen within vacuole sections of *atg7Δ* cells. Even in cells expressing Atg7, many vacuole sections had no visible bodies (left panels), likely due to the clustering of the bodies outside of the sectioning plane. However, on average fewer and smaller autophagic body cross-sections could be seen in vacuole sections from cells with lower levels of Atg7. Scale bars: 600 nm. (Figure adapted from ref. [18])

outlining in a blinded fashion, where the scientist outlining the images did not know the identity of each sample. When outlining vacuole sections, any bodies within the vacuole were included as part of the vacuole, and if a cell section contained more than one visible vacuole section, those vacuole sections were combined into one. When outlining body sections, each section was shifted slightly so that the outlines were not touching, as this made it easier to measure the outlines (Fig. 2a, b) (*see* **Note 5**).

Once all of the body sections and vacuole sections had been outlined, their areas were measured automatically using ImageJ (Fig. 2c). We made the simplifying assumption that all bodies and vacuoles were spherical, even though in fact the cross-sections were not perfect circles, as this made the measurements more amenable to analysis. Based on this assumption, the measured cross-sectional areas were converted into radii. If a collection of perfect spheres whose actual sizes follow a lognormal distribution are randomly

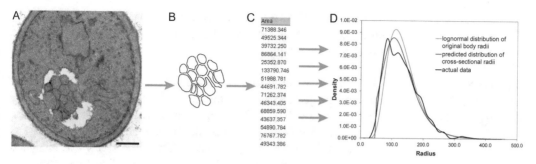

Fig. 2 Workflow for the quantification of autophagic body cross-sections and estimation of the size distribution of the original bodies. (**a, b**) Autophagic body cross-sections visible within a vacuole section were manually outlined in Adobe® Photoshop, with the addition of a small offset between each body outline to allow easier quantification. (**c**) Automated measurement of the area of each outlined body was performed in ImageJ. Areas were converted to radii by modeling each outline as a circle [$r = \sqrt{(A/\pi)}$]. (**d**) The distribution of radii from all cross-sections was used to fit a curve of expected cross-sections that would be generated by the sectioning of a lognormal distribution of original spherical bodies with tunable parameters. The resulting size distribution estimate and the number of observed cross-sections could also be used to estimate the original number of bodies

positioned in space and sliced with a random slice, we can estimate the original size distribution of the spheres from the distribution of the measured cross-sectional radii using established equations (Fig. 2d) [29]. Simulations using packed spheres verified that the clustering of the bodies does not significantly affect these results [20]. Excel sheets with built-in formulas to aid with estimating of the distribution of original body sizes and average number of bodies per vacuole are provided in the supplemental materials of the 2014 Backues et al. protocol paper [20].

Using these methods, multiple studies have estimated that wild-type cells produce approximately 10–14 autophagosomes per cell per hour during the first few hours of nitrogen starvation [13, 18, 20, 30].

3.3 Acquisition of Time-Lapse GFP-Atg8 Movies

Cells expressing GFP-Atg8 as their only copy of Atg8 were grown to log phase in SMD-complete medium (0.67% yeast nitrogen base w/o amino acids (ForMedium, CYN0401), 2% glucose, and auxotrophic amino acids and vitamins as needed; *see* **Note 6**) and starved for 30 min in SD-N to induce autophagy. Starved cells (100 μl) at an OD_{600} of 0.5–1.0 were pipetted onto the concanavalin A-coated coverslip, spread and allowed to adhere for 5 min at room temperature. The coverslip was rinsed five times by dipping into water to remove unbound cells, then immediately placed on top of a concavity slide filled with ~40 μl SD-N, with the cells in contact with the media. Slides were immediately mounted, coverslip down, onto an inverted deconvolution fluorescence microscope and live imaged using the GFP/FITC channel. Full z-stacks (15 images per stack, 0.4-μm spacing between stacks) were acquired at 1-min intervals for 45 min (*see* **Note** 7). Images were deconvolved and a maximum-intensity projection time series created using the software provided with the microscope.

3.4 Quantification of Time-Lapse GFP-Atg8 Movies

The output of the fluorescence microscopy data after deconvolution and projection was a time series of images, each separated by 1-min intervals (Fig. 3). Each image was a projection of all z-planes, so that all GFP-Atg8 puncta could be seen at once and not lost as they move between imaging planes. The time series was analyzed manually by creating a spreadsheet with a row for each cell in the image and a column for each timepoint. Each cell at each timepoint that contained a visible punctum was scored as having a punctum that was either brighter, dimmer, or the same intensity as the same punctum at the previous timepoint. An autophagosome formation event was identified as a GFP-Atg8 punctum that appeared, grew brighter, grew dimmer, then disappeared. We measured both the lifetime of the puncta, from its first appearance to its disappearance, and the frequency of puncta initiation. Any puncta whose entire lifetime (from appearance to disappearance) was not captured in the time series was omitted from analysis.

A typical cell might have a single punctum appear, grow brighter for 2–4 frames, stay the same for 0–3 frames, grow dimmer for 2–4 frames, and then disappear entirely, with another punctum appearing a couple of frames later. However, there was a large amount of variation from event to event, so that many did not conform perfectly to this "typical" pattern (Fig. 3; also *see* **Note 8**). It was also not uncommon to see more than one punctum in a given cell at a given time. Usually this was the result of an overlap of two autophagosome formation events, where a new puncta formed and began becoming brighter before an earlier puncta had entirely faded, but occasionally two puncta formed approximately concurrently.

Using this method, we have measured the frequency of autophagosome formation to be 7–9 autophagosomes per cell per hour in wild-type yeast during the early stages of nitrogen starvation [11] (and our unpublished data), (*see* **Note 9**).

4   Notes

1. It is critical that the GFP is attached to the N terminus of Atg8, not the C terminus, as the final amino acid at the Atg8 C terminus is proteolytically removed in the cytosol prior to the localization of Atg8 at the PAS [21, 31].

2. The culture must be very healthy prior to induction of starvation or many dead cells will be seen in the TEM images. Thus, we started a culture from a freshly struck plate never stored at

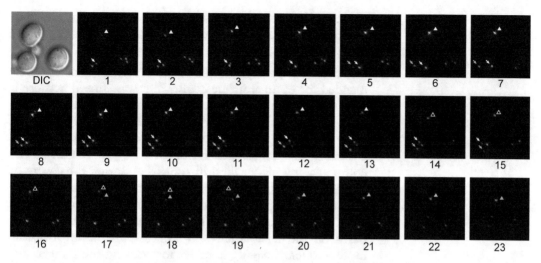

Fig. 3 Tracking individual rounds of autophagosome formation via GFP-Atg8 fluorescence imaging. Fluorescent z-stacks of cells undergoing nitrogen starvation were acquired every minute, deconvolved and used to generate a time series of maximum-intensity projections. The first 23 min of this time course are shown along with a differential interference contrast (DIC) image to show cell locations. GFP-Atg8-labeled puncta represent forming autophagosomes. Arrowheads (white, outlined, and gray) indicate three sequential autophagosome formation events (three rounds of puncta brightening and dimming) within the same cell. Arrows (white and gray) indicate two separate autophagosome formation events that overlap temporally within the same cell

4 °C, and maintained the culture in log phase for at least 24 h prior to starvation. Samples were also treated gently throughout the embedding process (e.g., no vortexing and only low-speed centrifugations) in order to avoid cell breakage.

3. Note that because the bodies cluster together there is a great amount of variation in the number of body sections visible in each cell section, just depending on whether or not that section happened to catch the cluster of bodies. In order to accurately estimate body number, it is critical to also include cells without visible bodies in the analysis, so do not omit those.

4. If your institution contains a TEM core, we recommend consulting with them before beginning this process as the staff there may be able to carry out most of these steps.

5. The manual outlining of autophagic bodies is the most time-consuming portion of the workflow. It may be possible to speed it up slightly by tweaking the automated methods for measuring the outlines so that the outlines do not need to be separated from each other but instead represent a segmentation of the autophagic body cluster.

6. Cells should be grown in an appropriate synthetic medium (such as SMD complete) prior to any sort of fluorescence imaging; use of YPD leads to bright autofluorescence that will mask the GFP-Atg8 signal. The addition of excess adenine

to the medium may help reduce autofluorescence by eliminating the accumulation of adenine precursors in *ade* mutant strains.

7. The chosen z-spacing (0.4-μm per slice) is twice what is recommended for optimal image quality, but was chosen because it gives sufficient resolution to track the puncta and allows the stack to be acquired in half the time. Fifteen slices per stack allow full coverage of essentially all of the cells adhered to the coverslip, including a little bit of extra margin. Note that the entire imaging plane will drift somewhat during the course of acquisition, making it necessary to watch the images as they are acquired and adjust the z-focus slightly, periodically, perhaps every 5–15 min. The timing of the imaging can be adjusted based on the capabilities of the microscope used; obviously more frequent imaging is better for tracking the puncta, but in practice the speed is limited by the need to acquire full z-stacks and to avoid excessive photobleaching over the entire 45–60-min time course.

8. If a single punctum partially faded and then brightened again, this was considered to be two separate events, on the assumption that the first punctum had actually disappeared and a second punctum appeared to replace it between frames; this occurred frequently.

9. TEM analyses of autophagic bodies have suggested slightly higher rates of autophagosome formation (10–14 per hour [18, 20, 30]) than determined by GFP-Atg8 tracking (7–9 per hour [11] and our unpublished data). This may simply reflect differences in experimental setup, such as the exact strain, growth conditions, and starvation time. Alternately, it may reflect biases in one or both of these methods. The TEM analysis contains a certain amount of subjectivity in the manual outlining of the autophagic bodies, which are sometimes difficult to distinguish, and error may also be introduced by the simplifying assumptions in the mathematical models used to estimate the size and number of the original 3D bodies from the collection of 2D cross-sections. The GFP-Atg8 tracking is complicated by the fact that there are sometimes multiple mobile puncta per cell and that frames are only collected once per minute so that occasionally the transition between events is unclear. Future improvements in both techniques, such as faster fluorescence imaging, improved mathematical modeling, and automated image processing, should continue to improve our confidence in these results. In the meantime, both techniques are already very suitable for the relative comparison of rates of autophagy in various mutant strains with those seen in wild-type strains processed in parallel, and have contributed important insights into the roles of Atg proteins and their regulators.

References

1. Eskelinen E-L, Reggiori F, Baba M et al (2011) Seeing is believing: the impact of electron microscopy on autophagy research. Autophagy 7:935–956. https://doi.org/10.4161/auto.7.9.15760

2. Takeshige K, Baba M, Tsuboi S et al (1992) Autophagy in yeast demonstrated with proteinase-deficient mutants and conditions for its induction. J Cell Biol 119(2):301–311

3. Baba M, Takeshige K, Baba N, Ohsumi Y (1994) Ultrastructural analysis of the autophagic process in yeast: detection of autophagosomes and their characterization. J Cell Biol 124:903–913

4. Baba M, Osumi M, Scott SV et al (1997) Two distinct pathways for targeting proteins from the cytoplasm to the vacuole/lysosome. J Cell Biol 139(7):1687–1695

5. He C, Song H, Yorimitsu T et al (2006) Recruitment of Atg9 to the preautophagosomal structure by Atg11 is essential for selective autophagy in budding yeast. J Cell Biol 175:925–935. https://doi.org/10.1083/jcb.200606084

6. Baba M, Tomonaga S, Suzuki M et al (2019) A nuclear membrane-derived structure associated with Atg8 is involved in the sequestration of selective cargo, the Cvt complex, during autophagosome formation in yeast. Autophagy 15:423–437. https://doi.org/10.1080/15548627.2018.1525475

7. Gómez-Sánchez R, Rose J, Guimarães R et al (2018) Atg9 establishes Atg2-dependent contact sites between the endoplasmic reticulum and phagophores. J Cell Biol 217:2743–2763. https://doi.org/10.1083/jcb.201710116

8. Noda T, Klionsky DJ (2008) Chapter 3. The quantitative Pho8Δ60 assay of nonspecific autophagy. Methods Enzymol 451:33–42. https://doi.org/10.1016/S0076-6879(08)03203-5

9. Guimaraes RS, Delorme-Axford E, Klionsky DJ, Reggiori F (2015) Assays for the biochemical and ultrastructural measurement of selective and nonselective types of autophagy in the yeast Saccharomyces cerevisiae. Methods 75:141–150. https://doi.org/10.1016/j.ymeth.2014.11.023

10. Torggler R, Papinski D, Kraft C (2017) Assays to monitor autophagy in Saccharomyces cerevisiae. Cell 6(3):23. https://doi.org/10.3390/cells6030023

11. Xie Z, Nair U, Klionsky DJ (2008) Atg8 controls phagophore expansion during autophagosome formation. Mol Biol Cell 19:3290–3298. https://doi.org/10.1091/mbc.E07-12-1292

12. Nakatogawa H, Ichimura Y, Ohsumi Y (2007) Atg8, a ubiquitin-like protein required for autophagosome formation, mediates membrane tethering and hemifusion. Cell 130:165–178. https://doi.org/10.1016/J.CELL.2007.05.021

13. Jin M, He D, Backues SK et al (2014) Transcriptional regulation by Pho23 modulates the frequency of autophagosome formation. Curr Biol 24:1314–1322. https://doi.org/10.1016/j.cub.2014.04.048

14. Legakis JE, Yen W-L, Klionsky DJ (2007) A cycling protein complex required for selective autophagy. Autophagy 3:422–432. https://doi.org/10.4161/auto.4129

15. Tucker KA, Reggiori F, Dunn WA, Klionsky DJ (2003) Atg23 is essential for the cytoplasm to vacuole targeting pathway and efficient autophagy but not pexophagy. J Biol Chem 278:48445–48452. https://doi.org/10.1074/jbc.M309238200

16. Yao Z, Delorme-Axford E, Backues SK, Klionsky DJ (2015) Atg41/Icy2 regulates autophagosome formation. Autophagy 11(12):2288–2299. https://doi.org/10.1080/15548627.2015.1107692

17. Feng Y, Backues SK, Baba M et al (2016) Phosphorylation of Atg9 regulates movement to the phagophore assembly site and the rate of autophagosome formation. Autophagy 12(4):648–658. https://doi.org/10.1080/15548627.2016.1157237

18. Cawthon H, Chakraborty R, Roberts JR, Backues SK (2018) Control of autophagosome size and number by Atg7. Biochem Biophys Res Commun 503:651–656. https://doi.org/10.1016/J.BBRC.2018.06.056

19. Xie Z, Nair U, Geng J et al (2009) Indirect estimation of the area density of Atg8 on the phagophore. Autophagy 5:217–220. https://doi.org/10.4161/auto.5.2.7201

20. Backues SK, Chen D, Ruan J et al (2014) Estimating the size and number of autophagic bodies by electron microscopy. Autophagy 10:155–164

21. Suzuki K, Kirisako T, Kamada Y et al (2001) The pre-autophagosomal structure organized by concerted functions of APG genes is essential for autophagosome formation. EMBO J 20:5971–5981. https://doi.org/10.1093/emboj/20.21.5971

22. Kim J, Huang W-P, Stromhaug PE, Klionsky DJ (2002) Convergence of multiple autophagy

and cytoplasm to vacuole targeting components to a perivacuolar membrane compartment prior to de novo vesicle formation. J Biol Chem 277:763–773. https://doi.org/10.1074/jbc.M109134200

23. Segarra VA, Boettner DR, Lemmon SK (2015) Atg27 tyrosine sorting motif is important for its trafficking and Atg9 localization. Traffic 16:365–378. https://doi.org/10.1111/tra.12253

24. Yamamoto H, Kakuta S, Watanabe TM et al (2012) Atg9 vesicles are an important membrane source during early steps of autophagosome formation. J Cell Biol 198:219–233. https://doi.org/10.1083/jcb.201202061

25. Backues SK, Orban DP, Bernard A et al (2015) Atg23 and Atg27 act at the early stages of Atg9 trafficking in S. cerevisiae. Traffic 16:172–190. https://doi.org/10.1111/tra.12240

26. Suzuki K, Kubota Y, Sekito T, Ohsumi Y (2007) Hierarchy of Atg proteins in pre-autophagosomal structure organization. Genes Cells 12:209–218. https://doi.org/10.1111/j.1365-2443.2007.01050.x

27. Reggiori F, Tucker KA, Stromhaug PE, Klionsky DJ (2004) The Atg1-Atg13 complex regulates Atg9 and Atg23 retrieval transport from the pre-autophagosomal structure. Dev Cell 6:79–90. https://doi.org/10.1016/S1534-5807(03)00402-7

28. Cheong H, Yorimitsu T, Reggiori F et al (2005) Atg17 regulates the magnitude of the autophagic response. Mol Biol Cell 16:3438–3453. https://doi.org/10.1091/mbc.E04-10-0894

29. Feuerverger A, Menzinger M, Atwood HL, Cooper RL (2000) Statistical methods for assessing the dimensions of synaptic vesicles in nerve terminals. J Neurosci Methods 103:181–190

30. Bernard A, Jin M, González-Rodríguez P et al (2015) Rph1/KDM4 mediates nutrient-limitation signaling that leads to the transcriptional induction of autophagy. Curr Biol 25:546–555. https://doi.org/10.1016/j.cub.2014.12.049

31. Klionsky DJ (2011) For the last time, it is GFP-Atg8, not Atg8-GFP (and the same goes for LC3). Autophagy 7:1093–1094. https://doi.org/10.4161/auto.7.10.15492

Chapter 2

Ultrastructure of the Macroautophagy Pathway in Mammalian Cells

Eeva-Liisa Eskelinen and Katri Kallio

Abstract

Autophagy was originally identified using transmission electron microscopy, and this technique is still one of the most sensitive approaches to detect and quantify the amount of autophagic structures in cells and tissues. However, identification of autophagosomes, amphisomes, and autolysosomes in transmission electron microscopy sections requires experience and caution. The purpose of this chapter is to provide guidelines for correct identification of autophagic structures by morphology in conventional transmission electron microscopy thin sections.

Keywords Phagophore, Isolation membrane, Autophagosome, Amphisome, Autolysosome, Autophagic vacuole, Electron microscopy

1 Introduction

In autophagy, cells transport their own cytoplasmic material and organelles to lysosomes for degradation and recycling [1–5]. Cytoplasmic material can be transported to lysosomes via at least three different routes, which are called macroautophagy, microautophagy, and chaperone-mediated autophagy. This chapter deals with the electron-microscopy morphology of the macroautophagy pathway in cultured mammalian cells, with applicability to neuronal cells.

The first step in macroautophagy is the engulfment of cytosolic cargo into membrane-bound *autophagosomes* (Fig. 1). Autophagosomes are formed by elongation, curving and closure of *phagophores*, also called *isolation membranes*. Phagophores are flat membrane cisterns that emerge next to the endoplasmic reticulum when autophagy initiates [6]. Because phagophores are membrane cisterns, autophagosomes have two lipid bilayers around them. Autophagosomes often fuse with multivesicular endosomes and form *amphisomes*. Finally, autophagosomes and/or amphisomes fuse with lysosomes, forming *autolysosomes*. Lysosomes deliver the

Ben Loos and Esther Wong (eds.), *Imaging and Quantifying Neuronal Autophagy*, Neuromethods, vol. 171,
https://doi.org/10.1007/978-1-0716-1589-8_2, © Springer Science+Business Media, LLC, part of Springer Nature 2022

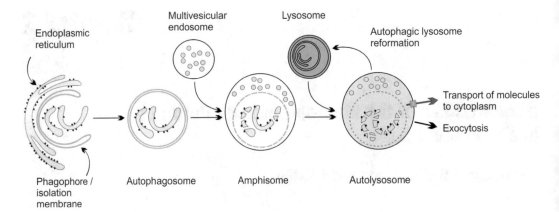

Fig. 1 Schematic drawing of the macroautophagy pathway in mammalian cells. In this example, the cytoplasmic cargo is pieces of rough endoplasmic reticulum. See text for further details on the cargo

majority of the enzymes that degrade the cytoplasmic cargo. Degradation products are transported back to cytoplasm by pumps located in the lysosomal membrane. The final stages of the autophagy pathway are less well known. It has been suggested that autolysosomes fuse with the plasma membrane and empty their contents to the extracellular space [7]. Another alternative is that lysosomes reform from autolysosomes [8]. These two alternatives are not mutually exclusive.

Autophagy was originally described using transmission electron microscopy in late 1950s, soon after this technology became available for cell biology researchers [9]. Several decades later, the first autophagy genes and proteins were discovered using yeast mutants [10], and this development has made it possible to use autophagy genes and proteins to manipulate and monitor the process in different model organisms, using a wide variety of genetic, biochemical, and fluorescence microscopy methods. Despite the fast increase of available new methods to monitor autophagy, electron microscopy still remains one of the important tools in autophagy research. One of its advantages is that no genetic or other type of manipulation of the cells is needed when electron microscopy is used to quantify the amount of autophagic structures, because autophagic structures can be identified by morphology. However, review of literature reveals that basic problems still exist in the correct identification of autophagic structures in transmission electron microscopy samples. Therefore, this chapter aims to provide simple guidelines to help in correct identification of autophagosomes, amphisomes, and autolysosomes by morphology in conventional transmission electron microscopy thin sections. The principles described here are also suitable to guide identification of autophagic structures in mammalian tissues, and at least to some extent, in cells and tissues of other species in addition to mammals. For the purpose of this series, the principles described here are aimed to guide identification of autophagic structures in neurons.

2   Materials

Cell and tissue samples prepared for transmission electron microscopy using conventional aldehyde fixation and plastic embedding protocols, or cryofixation followed by freeze substitution, are suitable for identification of autophagic structures by morphology. Since the purpose of this chapter is to concentrate on the identification of the organelles in electron microscopy thin sections, the sample preparation methods are not described here. Moreover, we have previously published detailed protocols for sample preparation [11–14], as well as for quantification of autophagic structures using electron microscopy [13].

The electron micrographs presented in this chapter were acquired using a Jeol 1200 EX transmission electron microscope. The majority of the micrographs were recorded using a bottom-mounted digital CCD camera. Some of the micrographs were taken using a bottom-mounted film camera. The photographic prints were scanned to make digital images.

3   Methods

How to identify phagophores, autophagosomes, amphisomes, and autolysosomes in thin sections?

3.1   Phagophores and Autophagosomes

In mammalian cells, autophagosomes are typically between 500 nm and 1.5 μm in diameter. Structures smaller than 500 nm in diameter may still be autophagosomes, but it is recommendable to use immunolabeling to confirm their identity. LC3 is a good marker protein for both phagophores and autophagosomes. When identifying phagophores and autophagosomes by morphology, attention should be paid *both* to the limiting membrane and to the contents, i.e., cytoplasmic cargo. Both *phagophores and autophagosomes should always have cytosolic contents*, which in most cases includes ribosomes, and very often also pieces of rough endoplasmic reticulum (Fig. 2). Occasionally, other cytoplasmic constituents like mitochondria or other organelles are visible among the cytoplasmic cargo. Many phagophores/autophagosomes also contain small vesicles or tubules (Fig. 2d, f, g–i) or filamentous material (Fig. 2i). The appearance of the double limiting membrane can vary. Part of the phagophores/autophagosomes show the two lipid bilayers clearly, since there is an empty cleft between the two membranes (Fig. 2a–e, j, l), while in part of the structures the two lipid bilayers are attached to each other and the limiting membrane resembles one thick membrane (Fig. 2f–i, k). In many cases, the limiting membrane can show the two membranes separated from each other in some areas, and attached to each other elsewhere (Fig. 2b, f, g, l). The same dual morphology of the double limiting

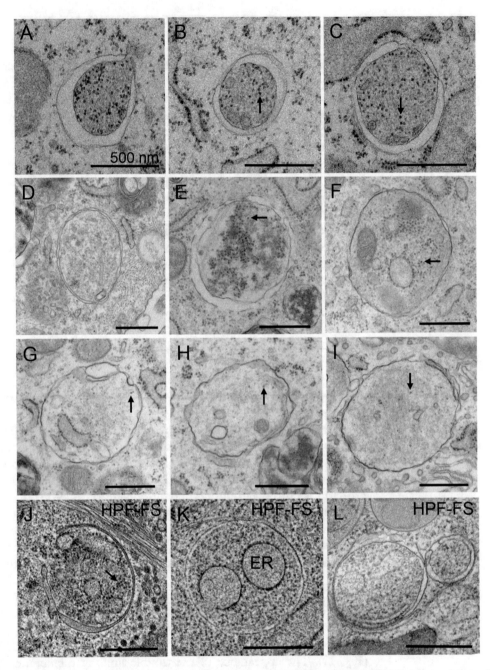

Fig. 2 Morphology of phagophores and autophagosomes in cultured mammalian fibroblast-like cell lines. (**a–i**) Conventional aldehyde fixation and epon embedding. (**j–l**) High-pressure freezing and freeze substitution (HPF-FS). Some of the ribosomes inside the autophagosomes are indicated by arrows. *ER* rough endoplasmic reticulum. Scale bars, 500 nm

membrane is visible both in conventional aldehyde-fixed, epon-embedded samples (Fig. 2a–i) and in high-pressure frozen, freeze-substituted samples (Fig. 2j–l). The latter method is considered to produce near-native morphology, since cryofixation immobilizes lipid membranes more faithfully than aldehyde fixation [11, 15] (*see* **Note 1**).

3.2 Amphisomes and Autolysosomes

The main criterion in the identification of amphisomes and especially autolysosomes is the cytoplasmic contents. Since the inner limiting membrane is degraded by the lysosomal enzymes together with the cytoplasmic cargo, autolysosomes may have only one limiting membrane. However, both amphisomes and autolysosomes may also still have two limiting membranes visible (Fig. 3a, b, i). Ribosomes are a good feature as a marker of the cytoplasmic cargo also in the degradative autophagic structures. Acidification of the contents and/or action of the lysosomal enzymes turn ribosomes into a characteristic electron-dense mass (Fig. 3b–i). Mitochondria or other organelles can also be visible among the contents. In many cases, the degradative autophagic structures show several internal compartments, produced by fusion of several autophagosomes with each other and/or with the same endosome or lysosome (Fig. 3e, f). In these structures, the cytoplasmic cargo is still surrounded by the inner limiting membrane of the autophagosome. Typically, a degradative autophagic structure contains autophagic cargo surrounded by the inner limiting membrane, small internal vesicles typical for multivesicular endosomes (Fig. 3a–c) and/or finger-print-like multilamellar membranes typical for lysosomes (Fig. 3a, c–e, h). However, it is important to note that the presence of small internal vesicles or finger-print like multilamellar membranes *without* the cytoplasmic cargo does not mean that the structure in question is a degradative autophagic compartment (*see* **Notes 2–4**).

4 Notes

1. *Phagophore, autophagosome, or amphisome?*

   Conventional thin sections are 60–80 nm thick, while the diameter of autophagosomes is between 500 nm and 1.5 μm. Thus, one thin section gives a two-dimensional image of the three-dimensional objects. One thin section can slice the structure in a random orientation. This means that in a thin section, a phagophore can look similar to a sealed autophagosome. In most cases this is not a problem, since both phagophores and autophagosomes represent early macroautophagic structures and their presence indicates autophagosome formation. However, if it is important to differentiate phagophores and autophagosomes, more advanced identification methods need to be employed. These can be e.g., immunolabeling with specific marker protein antibodies, or three-dimensional electron microscopy, which however is not well applicable to

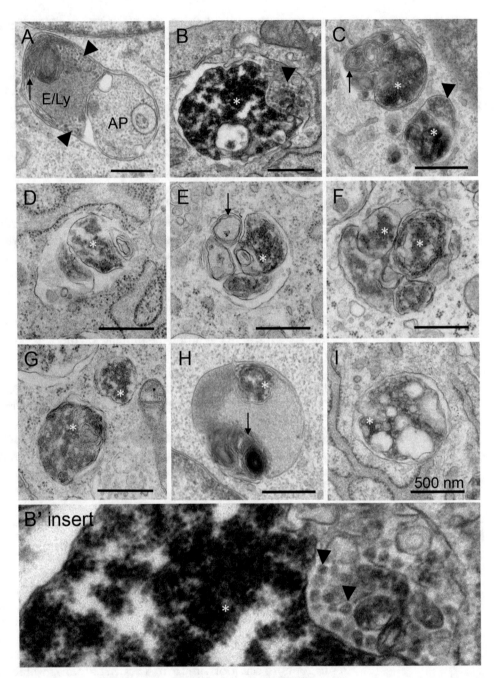

Fig. 3 Morphology of amphisomes/autolysosomes in cultured mammalian fibroblast-like cell lines (**a–i**). Asterisks indicate electron-dense, partially degraded ribosome aggregates. Small internal vesicles delivered by fusion of multivesicular endosomes are indicated by arrowheads. Arrows indicate multilamellar membranes typically found inside lysosomes. Panel (**b′**) insert shows a higher magnification of a region of the same amphisome as in panel (**b**), to demonstrate the partially degraded ribosomes and the small internal vesicles. *E/Ly* late endososome or lysosome; *AP* autophagosome. All images are from samples prepared using conventional aldehyde fixation and epon embedding. Scale bars, 500 nm

quantitative approaches due to its time-consuming and work-intensive nature. It is also possible that in a thin section a structure has the morphology of an autophagosome, although it has already fused with an endosome or even a lysosome. As an example, the structure in Fig. 3a could give such impression if the thin section would be oriented differently. To be on the safe side, many researchers prefer to use the term *initial autophagic vacuole* (AVi) to describe structures that look like autophagosomes in thin sections, but could actually be either phagophores, autophagosomes, or amphisomes.

2. *Amphisome or autolysosome?*

In most cases, it is not possible to unambiguously identify whether a degradative autophagic structure, i.e., an autophagic vacuole containing partially degraded cytoplasmic cargo, is an amphisome or an autolysosome. Both show signs of degradation among the cytoplasmic cargo, and both may contain small internal vesicles delivered by fusion of a multivesicular endosome. However, the presence of multilamellar membranes among the contents, together with the partially degraded cytoplasmic cargo, suggests that the structure has fused with a lysosome, and thus is likely to represent an autolysosome. The term *degradative autophagic vacuole* (AVd) refers to both amphisomes and autolysosomes.

3. *Empty vacuoles and swollen mitochondria are not autophagic structures.*

The most common organelles incorrectly identified as autophagic structures are empty cytoplasmic vacuoles and swollen mitochondria. Cytoplasmic vacuoles are limited by one lipid bilayer, and they have no or very little visible contents (Fig. 4a–c). Swollen mitochondria can sometimes be misleading, since like autophagosomes, mitochondria have two limiting membranes. Especially when mitochondria are swollen, it is possible that a thin section of a certain mitochondrion does not show any cristae, or the cristae may be difficult to identify due to unusual morphology (Fig. 4d–f). As a guideline, the two limiting membranes of mitochondria are very close to each other, and the distance between the two lipid bilayers is relatively constant compared to the typically variable distance between the two autophagosome limiting membranes. Although mitochondria have their own ribosomes, these are very rarely visible in electron micrographs. On the contrary, autophagosomes almost always have ribosomes visible among their cytoplasmic cargo.

4. *Other pitfalls in identification of autophagic structures.*

It should be emphasized that autophagosome limiting membranes are devoid of attached ribosomes. In thin sections, rough endoplasmic reticulum can occasionally look like a

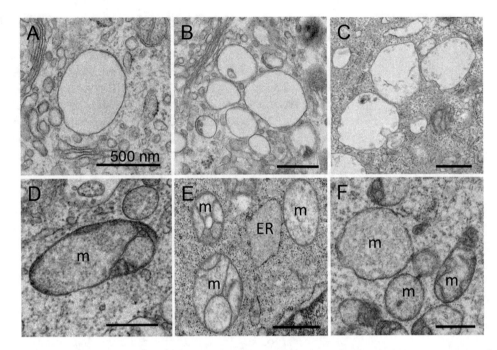

Fig. 4 Empty cytoplasmic vacuoles (**a–c**) and swollen mitochondria (**d–f**). These structures are sometimes incorrectly identified as autophagic vacuoles. Note that similar to autophagosomes, also mitochondria have two limiting membranes, and cristae are not always visible in thin sections (panel **f**). *m* mitochondrion; *ER* lumen of rough endoplasmic reticulum. All images are from samples prepared using conventional aldehyde fixation and epon embedding. Scale bars, 500 nm

vesicle surrounded by a double membrane and containing cytoplasm [16]. However, the presence of attached ribosomes makes it easy to identify rough endoplasmic reticulum and to avoid this misinterpretation. Another mistake that has occurred in literature is the confusion of a lipid bilayer and a double membrane [17]. This incorrect interpretation leads to identification of early endosomes as autophagic vacuoles. The additional mistake in this case is that early endosomes lack the cytoplasmic cargo, which should always be present in both initial and degradative autophagic structures.

Acknowledgments

The authors are supported by the Academy of Finland. The Electron Microscopy Unit at the Institute of Biotechnology, University of Helsinki, and the Electron Microscopy Laboratory at the Institute of Biomedicine, University of Turku, are thanked for technical help and availability of instruments. We also thank Helena Vihinen, Institute of Biotechnology, University of Helsinki, for help in high-pressure freezing and freeze substitution.

References

1. Kaushik S, Cuervo AM (2018) The coming of age of chaperone-mediated autophagy. Nat Rev Mol Cell Biol 19(6):365–381. https://doi.org/10.1038/s41580-018-0001-6

2. Tekirdag K, Cuervo AM (2018) Chaperone-mediated autophagy and endosomal microautophagy: joint by a chaperone. J Biol Chem 293(15):5414–5424. https://doi.org/10.1074/jbc.R117.818237

3. Mercer TJ, Gubas A, Tooze SA (2018) A molecular perspective of mammalian autophagosome biogenesis. J Biol Chem 293(15):5386–5395. https://doi.org/10.1074/jbc.R117.810366

4. Galluzzi L, Baehrecke EH, Ballabio A, Boya P, Bravo-San Pedro JM, Cecconi F, Choi AM, Chu CT, Codogno P, Colombo MI, Cuervo AM, Debnath J, Deretic V, Dikic I, Eskelinen EL, Fimia GM, Fulda S, Gewirtz DA, Green DR, Hansen M, Harper JW, Jaattela M, Johansen T, Juhasz G, Kimmelman AC, Kraft C, Ktistakis NT, Kumar S, Levine B, Lopez-Otin C, Madeo F, Martens S, Martinez J, Melendez A, Mizushima N, Munz C, Murphy LO, Penninger JM, Piacentini M, Reggiori F, Rubinsztein DC, Ryan KM, Santambrogio L, Scorrano L, Simon AK, Simon HU, Simonsen A, Tavernarakis N, Tooze SA, Yoshimori T, Yuan J, Yue Z, Zhong Q, Kroemer G (2017) Molecular definitions of autophagy and related processes. EMBO J 36(13):1811–1836. https://doi.org/10.15252/embj.201796697

5. Damme M, Suntio T, Saftig P, Eskelinen EL (2015) Autophagy in neuronal cells: general principles and physiological and pathological functions. Acta Neuropathol 129(3):337–362. https://doi.org/10.1007/s00401-014-1361-4

6. Axe EL, Walker SA, Manifava M, Chandra P, Roderick HL, Habermann A, Griffiths G, Ktistakis NT (2008) Autophagosome formation from membrane compartments enriched in phosphatidylinositol 3-phosphate and dynamically connected to the endoplasmic reticulum. J Cell Biol 182(4):685–701

7. Medina DL, Fraldi A, Bouche V, Annunziata F, Mansueto G, Spampanato C, Puri C, Pignata A, Martina JA, Sardiello M, Palmieri M, Polishchuk R, Puertollano R, Ballabio A (2011) Transcriptional activation of lysosomal exocytosis promotes cellular clearance. Dev Cell 21(3):421–430. https://doi.org/10.1016/j.devcel.2011.07.016

8. Yu L, McPhee CK, Zheng L, Mardones GA, Rong Y, Peng J, Mi N, Zhao Y, Liu Z, Wan F, Hailey DW, Oorschot V, Klumperman J, Baehrecke EH, Lenardo MJ (2010) Termination of autophagy and reformation of lysosomes regulated by mTOR. Nature 465(7300):942–946. https://doi.org/10.1038/nature09076

9. Eskelinen EL, Reggiori F, Baba M, Kovacs AL, Seglen PO (2011) Seeing is believing: the impact of electron microscopy on autophagy research. Autophagy 7(9):935–956

10. Tsukada M, Ohsumi Y (1993) Isolation and characterization of autophagy-defective mutants of Saccharomyces cerevisiae. FEBS Lett 333(1–2):169–174

11. Biazik J, Vihinen H, Jokitalo E, Eskelinen EL (2017) Ultrastructural characterization of phagophores using electron tomography on cryoimmobilized and freeze substituted samples. Methods Enzymol 587:331–349. https://doi.org/10.1016/bs.mie.2016.09.063

12. Biazik JM, Vihinen H, Anwar T, Jokitalo E, Eskelinen EL (2015) The versatile electron microscope: an ultrastructural overview of autophagy. Methods 75:44–53. https://doi.org/10.1016/j.ymeth.2014.11.013

13. Yla-Anttila P, Vihinen H, Jokitalo E, Eskelinen EL (2009) Monitoring autophagy by electron microscopy in mammalian cells. Methods Enzymol 452:143–164

14. Eskelinen EL (2008) Fine structure of the autophagosome. In: Deretic V (ed) Methods in molecular biology, vol 445: autophagosome and phagosome. Humana Press, Totowa, NJ, pp 11–28

15. Dubochet J (1995) High-pressure freezing for cryoelectron microscopy. Trends Cell Biol 5(9):366–368

16. Eskelinen EL (2008) To be or not to be? Examples of incorrect identification of autophagic compartments in conventional transmission electron microscopy of mammalian cells. Autophagy 4:257–260

17. Eskelinen EL, Kovacs AL (2011) Double membranes vs. lipid bilayers, and their significance for correct identification of macroautophagic structures. Autophagy 7(9):931–932

Chapter 3

Live Imaging of Autophagosome Biogenesis and Maturation in Primary Neurons

Andrea K. H. Stavoe and Erika L. F. Holzbaur

Abstract

Live-cell imaging of autophagy in primary neurons has revealed a robust and constitutive pathway for nonselective autophagy in the axon. Autophagosome biogenesis occurs in the distal axon; newly formed autophagosomes engulf cargos including mitochondrial fragments, protein aggregates, and bulk cytoplasm. Once formed, autophagosomes move rapidly and processively along the axon toward the soma, fusing with lysosomes in transit to mature into degradation-competent autolysosomes. Each step of the biogenesis and maturation pathway can be visualized with live imaging of primary neurons in culture; importantly, live imaging in vivo in *C. elegans* and *Drosophila* has confirmed observations made in primary neurons. Here, we detail considerations relating to choice of model system, probe, and microscope, and provide advice on methods and specific points to consider.

Keywords Autophagy, Autophagosome biogenesis, Autophagosome maturation, Axonal transport, Mitophagy, Live-cell imaging, Primary neurons, Hippocampal neurons, Dorsal root ganglion neurons

1 Introduction

1.1 Dynamics of Autophagosome Biogenesis and Maturation in Neurons

The autophagy pathway has been well characterized in yeast, as recognized by the award of the Nobel Prize in Physiology or Medicine to Dr. Yoshinori Ohsumi in 2016. There is remarkable evolutionary conservation of the pathway, as most of the components identified in yeast function in a similar manner in mammalian cells (described by Klionsky elsewhere in this volume). What is less understood is how the stress-induced pathway of autophagy in yeast is adapted to specific functions in fully differentiated cells from higher eukaryotes such as neurons. Landmark studies have shown that inhibition of autophagy in neurons is sufficient to induce neurodegeneration [1, 2], emphasizing the importance of this pathway for neuronal health and function. However, we do not yet fully understand the dynamics of autophagy in neurons, how the pathway is regulated either temporally or spatially, or how and

Ben Loos and Esther Wong (eds.), *Imaging and Quantifying Neuronal Autophagy*, Neuromethods, vol. 171,
https://doi.org/10.1007/978-1-0716-1589-8_3, © Springer Science+Business Media, LLC, part of Springer Nature 2022

why disruption of autophagy is consistently linked to the development of neurodegenerative disease.

1.1.1 Cell Biology of the Neuron

Neurons are highly polarized cells, with distinct somal, axonal, and dendritic compartments. These compartments can differ significantly in their organelle complement and cytoskeletal organization. For example, while the endoplasmic reticulum (ER) is found throughout the neuron [3], Golgi bodies are found within the soma and dendrites, but are generally restricted from the axon [4]. Organelles in the degradative pathway, such as late endosomes and lysosomes, also show compartment-specific differences in localization. Although late endosomes and lysosomes are found throughout the neuron, lysosomes localized to the axon are less acidified and have a more limited complement of degradative enzymes than those within the soma [5].

One underlying cause for these differences is the distinct cytoskeletal organization of neurons. Microtubules, which are the major tracks for the molecular motor-driven long-distance transport of organelles, are differentially organized in axons versus dendrites. In axons, microtubules are found uniformly organized with their "plus ends" orientated away from the soma. This organization defines the overall polarity for organelle transport within this compartment as the molecular motors, cytoplasmic dynein and kinesin, move unidirectionally in opposing directions along microtubules. In contrast, microtubule polarity is mixed within the dendrites of mammalian neurons, and a distinct set of motors drive organelle trafficking [6]. Further, trafficking of organelles and vesicles from the soma into the axon is tightly regulated by molecular filters at or near the axon initial segment (AIS), which discriminate between axonal versus dendritic cargos [7, 8].

Together, these mechanisms ensure that the soma, the axon, and dendrites have distinct complements of organelles and distinct patterns of organelle trafficking in both the biosynthetic and degradative pathways. Importantly, studies to date both in primary neurons and in vivo indicate that the autophagy pathway is differentially regulated in the soma, axons, and dendrites [9].

1.1.2 Imaging Autophagy in the Neuron

Given the spatial specificity of organelle trafficking in neurons, biochemical approaches that homogenize cellular contents do not provide the resolution required to understand the mechanisms leading to generation and maturation of autophagosomes in these highly polarized cells. Approaches that retain spatial specificity, such as light or electron microscopy, are required. Further, evidence to date indicates that regulation of autophagy is cell- and tissue-type specific [10], so methods that can clearly differentiate between autophagy in neurons versus autophagy in surrounding cell types (glia or muscle) are required to provide cell-type specific insights.

Autophagy has often been studied in the context of fixed cells, immunostained for autophagy components. While immunocytochemical approaches allow for the interrogation of endogenous proteins, fixed cells cannot offer insights into the kinetics of the autophagy pathway such as rates of formation, maturation, or flux.

For these reasons, we and others [4, 11–14] have used live-cell imaging of primary neurons to study autophagosome biogenesis, cargo-loading, and organelle maturation, as well as the upstream regulatory pathways and downstream consequences of autophagic function. Here we discuss the methods we have used to image the dynamics of autophagy in primary neurons. We highlight the strengths and limitations of this approach, and discuss considerations of model system, probe and microscope selection, data collection, and data analysis. As we are now in an age of rapid development of ever more powerful microscopes, synergizing with the discovery of increasingly bright and stable fluorescent probes, we predict that these approaches will continue to improve, providing important insights into our understanding of autophagy in neurons.

1.1.3 Conserved Pathway for Constitutive Autophagosome Biogenesis and Maturation in the Axon

Imaging autophagosome biogenesis and maturation in primary neurons [4, 12, 13, 15, 16] and in vivo [17–19] has identified a conserved pathway for constitutive axonal autophagy in neurons (Fig. 1). Neurons exhibit the ordered recruitment of autophagy components during autophagosome biogenesis that was first elucidated in non-neuronal mammalian cells [20, 21]. Autophagosomes are preferentially generated in the distal axon; a new autophagosome is generated every 3–9 min in primary neurons in culture

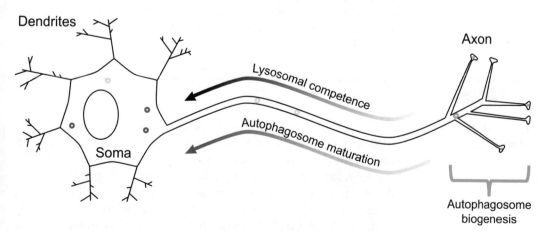

Fig. 1 Spatial organization of autophagy in neurons. Schematic illustrating the localized biogenesis of autophagosomes in the distal axon, followed by the translocation of these organelles toward the soma. Lysosomal competence in the axon decreases with increasing distance from the soma. Autophagosomes mature as they move from the distal axon to the proximal axon and soma via retrograde transport by fusing with lysosomes en route

[4, 13] and every 30 min in *C. elegans* AIY interneurons in vivo [17]. Once formed, these autophagosomes bind to microtubule-based motors. Initially, autophagosomes move bidirectionally, driven by both the microtubule-plus-end directed kinesin-1 and the microtubule-minus-end directed cytoplasmic dynein motor proteins [16]. Recruitment of the scaffolding protein JIP1 to the autophagosome inhibits kinesin-1, allowing the unopposed activity of cytoplasmic dynein to drive the processive movement of these organelles back to the soma [16, 18]. In transit, autophagosomes fuse with lysosomes, leading to acidification of the resulting auto-lysosome and the acquisition of degradative enzymes that digest internalized contents [13, 15].

This axonal autophagy pathway occurs constitutively and is not substantially upregulated by starvation [22] or expression of aggregation-prone proteins [13, 15] in primary neurons in vitro. These observations are consistent with previous work demonstrating that starvation does not significantly increase the level of autophagy observed in the brain in vivo [10].

While there is robust autophagy in the axon, live imaging of primary neurons reveals surprisingly low levels of constitutive autophagosome biogenesis in the dendrites or the soma under resting conditions [22]. However, autophagy has been shown to be required for developmental pruning of dendritic spines [23] and may respond to changes in neuronal activity. In contrast to the prominent role of nonselective autophagy characterized in axons, studies in *Drosophila* report that at least one form of selective autophagy, the PINK1- and Parkin-dependent degradation of damaged mitochondria via mitophagy, primarily occurs in the soma [24, 25]. However, Parkin-dependent mitophagy has also been observed in the axons of vertebrate neurons [26] so more work is required to establish the primary sites of mitochondrial quality control in the neuron [27].

2   Materials

2.1   Selection of Model System

A number of cell lines have been used over the years to model organelle dynamics in neurons, such as HeLa and PC12 cells. In our hands, however, none of these cell lines faithfully model the stereotypical cytoskeletal organization and organelle trafficking dynamics observed in primary neurons or in vivo. In contrast, numerous studies have shown that primary neurons model many of the trafficking pathways seen in neurons in vivo, in terms of the organization of the cytoskeleton, the selectivity of the AIS, and the transport of cargos along the axon and within dendrites [6]. These cell lines also do not model the robust and constitutive formation of autophagosomes observed in axons both in vitro and in vivo.

We have extensive experience with the imaging of organelle dynamics in dorsal root ganglion (DRG), hippocampal, and cortical neurons isolated from either rat or mouse. All three are well-characterized model systems, with some distinct advantages and disadvantages, discussed below. Importantly, key observations on autophagosome dynamics using all three cell types have been consistent, and importantly, have been replicated in vivo. Isolating primary neurons from rodents allows the use of transgenic mouse lines such as the GFP-LC3B mouse [10]. This approach ensures that autophagy is not disrupted by LC3B overexpression, which can sometimes be an issue at higher expression levels [4, 13].

iPSC-derived neurons are an increasingly popular model for the study of neuronal cell biology [28], including live-cell imaging of organelle dynamics. While the overall organization and intracellular dynamics of these cells have not been mapped with the same degree of specificity as for hippocampal neurons, this situation is changing very rapidly as researchers take advantage of iPSCs to study autophagy and mitophagy [29]. There have also been elegant studies of the dynamics of autophagy biogenesis and maturation in vivo, primarily in *C. elegans* and *Drosophila* [17–19, 30]. This work has correlated very well with observations made from primary neurons, leading to a coherent view of axonal autophagy in neurons.

2.2 Isolation and Culture of Primary Rodent Neurons

2.2.1 Selection of Neuronal Type

Live imaging of autophagy has been performed in hippocampal, cortical, and DRG neurons from rodents, both mouse and rat. The growth and differentiation of hippocampal neurons has been well characterized, with defined stages relating to axon specification, axon and dendrite outgrowth, and synaptic development [31], making this a powerful model system. Cortical neurons are somewhat easier to isolate and more plentiful, but the resulting cultures can be more heterogeneous than hippocampal cultures. Both of these cell types develop distinct axonal and dendritic compartments defined by distinct cytoskeletal and organelle markers.

DRG neurons do not develop distinct axonal and dendritic processes, but do extend neurites that are axon-like in the organization of their microtubule cytoskeleton and overall distribution of organelles. DRG neurons exhibit robust constitutive axonal autophagy and have proven to be a strong and highly predictive model for neuronal autophagy in vivo [13]. Further, DRG neurons have a significant advantage over hippocampal or cortical neurons, as this neuronal subtype can be readily isolated and successfully cultured from mice of any age [32]. Once plated as dissociated cultures, both embryonic and adult DRG neurons exhibit rapid outgrowth of neurites and robust organelle dynamics.

2.2.2 Hippocampal and Cortical neurons

Primary neurons can be readily isolated from the hippocampus or cortex of embryonic mice and rats, as previously described [31, 33]. Once isolated, hippocampal neurons can be plated onto glass-bottomed 35 mm P35G-1.5-20-C dishes (MatTek) coated with 500 µg/ml Poly-L-Lysine (PLL). After the PLL is washed off and rinsed with water, plates are allowed to dry for 5 min in a biosafety cabinet. Attachment media (MEM (Gibco) media supplemented with 10% heat-inactivated horse serum (Gibco), 1 mM pyruvic acid (Gibco), and 33 mM glucose (Sigma)) is added to plates and allowed to equilibrate at 37 °C and 5% CO_2. Neurons are plated into attachment media at a density of ~300 neurons per mm^2. After approximately 4 h, attachment media is removed and replaced with neurobasal (Gibco) media supplemented with 2% B-27 (Gibco), 33 mM glucose (Sigma), 2 mM GlutaMAX (Gibco), 100 U/mL penicillin (Sigma), and 100 mg/ML streptomycin (Sigma); neurons can be cultured for 1–2 weeks. Alternatively, neurons can be plated onto a glass coverslip compatible with insertion into an imaging chamber. We add the inhibitor cytosine β-D-arabinofuranoside (AraC; Sigma) at 1 µM on DIV 2 to prevent proliferation of glial cells in the culture. Media should be partially replaced (exchanging no more than 30–50% of total volume) every 3–4 days.

2.2.3 DRG Neurons

We dissect and isolate DRG neurons as described [34]. We transfect DRG neurons by electroporation prior to plating (*see* below) to express constructs of interest. To prepare the dishes, glass-bottomed 35 mm P35G-1.5-14-C dishes (MatTek) are coated with 0.01% PLL (Sigma) and incubated overnight at 4 °C. Plates are then washed twice with water and allowed to dry for 5 min in a biosafety cabinet. Dishes are then coated with 2 µg/mL Laminin (Corning) and incubated at 4 °C for at least 2 h. Prior to neuron plating, the laminin is removed and plates are equilibrated at 37 °C and 5% CO_2 in F-12 Ham's media (Invitrogen) supplemented with 10% heat-inactivated fetal bovine serum (HyClone), 100 U/mL penicillin, and 100 µg/mL streptomycin. DRG neurons are cultured for 2 days in complete F-12 media. Cultures are typically imaged on DIV 2.

2.3 Transfecting Neurons

2.3.1 Hippocampal and Cortical Neurons

We usually transfect hippocampal and cortical neurons 24–48 h prior to imaging, using Lipofectamine 2000 (Invitrogen), according to the manufacturer's instructions. We transfect DNA plasmids encoding fluorescently tagged constructs 24 h prior to imaging. When introducing siRNA, we use Lipofectamine RNAiMAX (Invitrogen) 48 h prior to imaging. Importantly, if we are introducing both siRNA and DNA plasmids, we transfect all constructs simultaneously using Lipofectamine 2000; Lipofectamine RNAiMAX will not successfully transfect DNA plasmids. In our hands, neurons do not survive two rounds of transfection. Using Lipofectamine 2000, we transfect in a maximum of 1.5 µg DNA per plate. Typically, 400 ng of each DNA construct is used. For longer term cultures, we transfect using a DNA-calcium phosphate co-precipitation approach on DIV 7 [35].

2.3.2 DRG Neurons

Following isolation but prior to plating, DRG neurons are resuspended in Isotonic Nucleofection Buffer (INB) solution (135 mM KCl, 0.2 mM $CaCl_2$, 2 mM $MgCl_2$, 10 mM HEPES pH 7.3, 5 mM EGTA). DRG neurons are then electroporated using an Amaxa Nucleofector (Lonza) according to manufacturer's directions using SCN (small cell number) Basic Neuron program #6, then plated in glass-bottomed dishes. Neurons are maintained for 2 days at 37 °C in a 5% CO_2 incubator. Up to 3 DNA constructs can be simultaneously electroporated into DRG neurons. For each plate, we electroporate at least 20,000 neurons with a maximum of 0.6 µg of total DNA (all constructs put together) per plate. Typically, only two constructs are simultaneously electroporated, with 250 ng of each construct.

2.4 Preparation of Neurons Prior to Imaging

Imaging neurons soon after plating will allow observations of the role of autophagy in neurite outgrowth (DIV 1 for DRG neurons and DIV 1–3 for hippocampal neurons, for example). Alternatively, imaging hippocampal cultures 7 days in vitro (7 DIV) or longer allows the observation of autophagy in more fully differentiated neurons with distinct axonal and dendritic processes. Imaging cultures 2–3 weeks after plating allows the analysis of autophagy in neurons that are synaptically connected; activity can be either induced or inhibited to study effects on biogenesis or flux (*see* **Note 1**).

2.4.1 Labeling with Fluorescent Ligands

Neurons transfected with Halo-tagged or SNAP-tagged constructs need to be incubated with the relevant fluorescent ligands. Cell-permeant ligands can be used to label intracellular proteins (SNAP-Cell ligands from NEB; HaloTag ligands from Promega), which enter the cell and bind irreversibly to SNAP- or Halo-tagged fusion proteins. We usually utilize the red (TMR) and far-red ligands with our SNAP- and Halo-labeled constructs, leaving the green (488 nm) channel for a GFP-tagged construct.

For hippocampal and cortical neurons, half of the media is removed from the plate and temporarily stored (kept warm and equilibrated). Ligand is then added to the remaining media and incubated at 37 °C for 30 min. Neurons are quickly and gently washed with pre-equilibrated complete media twice. For the third wash, the previously kept media is supplemented 1:1 with pre-equilibrated fresh media and then added to the plates. Neurons remain in the third wash for at least 15 min. Prior to imaging, the culture media is switched to Hibernate E, a low fluorescence imaging medium from BrainBits, supplemented with 2% B-27 and 2 mM GlutaMAX.

For DRG neurons, ligand is added directly to the neurons and incubated at 37 °C for 30 min. Plates are then washed three times with pre-equilibrated complete F-12 media. The third wash is left on the plates for at least 15 min at 37 °C. Prior to imaging, the culture media is switched to Hibernate A, a low fluorescence imaging medium from BrainBits, supplemented with 2% B-27 and 2 mM GlutaMAX.

Vital dyes such as Lysotracker or Mitotracker can also be added to the culture dish just prior to imaging, in the Hibernate medium.

2.5 Microscope Selection

Breakthrough work by Hollenbeck [11] used phase contrast microscopy, as well as epifluorescence microscopy, to image autophagosomes in the axons of primary neurons. We have used epifluorescence to monitor autophagosome dynamics in neurites [13, 15]. Given the much greater depth of the soma and even dendrites, autophagosomes in these compartments are not well resolved by epifluorescence and instead require confocal microscopy, as described below.

To analyze axonal dynamics, a research-grade epifluorescence microscope is required, equipped with a 60× or 100× high numerical aperture (NA) objective and a high-quality camera such as an EM-CCD or CMOS device. It is essential that the cultures be maintained at 37 °C during imaging. This can be accomplished using an isolation chamber that surrounds the culture dish such as a Chamlide CMB magnetic imaging chamber (BioVision Technologies), although we prefer to use a microscope equipped with a larger environmental chamber that fully encompasses the stage. This design allows the use of glass-bottom culture dishes and speeds the imaging of multiple cultures in a single imaging session. Because the microscope itself is a heat sink, we find it essential to pre-warm the environmental chamber for 30–60 min prior to imaging.

While epifluorescence offers sufficient resolution to view autophagosome dynamics in neurites, we strongly recommend the use of a spinning disk confocal (SDC) microscope to image all compartments of the neuron. Additionally, GFP-LC3 is typically easier to visualize on a SDC microscope, as the improved contrast allows detection of partially quenched GFP-LC3-labeled autophagosomes. We use a Perkin-Elmer UltraView Vox Spinning Disk Confocal system with a Yokogawa head on a Nikon Eclipse Ti Inverted Microscope, equipped with a Perfect Focus System, but any analogous spinning disk system should work. Our system utilizes a Hamamatsu EM-CCD camera, but newer CMOS cameras should also be considered. Fluorophores are excited using 405, 488, 561, and 640 nm laser lines; we have a dichroic mirror corresponding to those excitation wavelengths and emission wavelengths of 416–475, 500–549, 574–626, and 659–750 nm, respectively. This setup allows us to monitor up to four fluorescently tagged constructs in the same experiment, and cells are imaged with either a 60×/1.4 NA Plan Apo or an Apochromat 100×/1.49 NA oil-immersion objective from Nikon.

2.6 Selection of Fluorescent Probes to Image

The best-characterized marker for monitoring autophagy is LC3B, usually imaged as a fusion with GFP. This marker is cytoplasmic in cells prior to lipidation of LC3B and incorporation into a growing autophagosome. Once incorporated, it remains stably associated with the autophagosome until it is either acid-quenched or degraded. In contrast, expression of a mCherry-tagged LC3B construct will allow imaging of both the initially formed autophagosome and the mature and acidified autolysosome, as this fluorophore is not quenched in the acidic environment. Many investigators have taken advantage of this difference by expressing a tandem mCherry-EGFP-LC3B construct [36], so that both immature, unacidified autophagosomes and acidified autolysosomes can be identified by the presence of red fluorescence with or without green signal and used as a proxy for autophagic flux.

A caveat to this approach is that the overexpression of GFP-LC3B, or any marker, can potentially alter the dynamics of the pathway. One way to avoid this issue is to isolate neurons from a transgenic mouse line expressing GFP-LC3 (line #53) expressed in C57BL/6JJcl mice, developed by Mizushima et al. [10] and available from the RIKEN BioResource Center in Japan. These mice express low but consistent levels of the GFP-LC3B transgene in neurons without a detectable phenotype, providing assurance that the expression levels observed in neurons do not disrupt the processes under study.

To monitor upstream steps in autophagosome biogenesis, we have used fluorescently tagged constructs of ATG13, ATG5, DFCP1, WIPI2B, ATG14, ATG2A, and ATG9 (available from Addgene); for example, *see* [4, 32]. Importantly, tagged ATG16L1 should not be transfected into cells, as overexpression of ATG16L1 has been shown to interfere with autophagosome biogenesis [37]. Autophagosome markers require special attention when selecting fluorescent tags, as most autophagy proteins exist diffusely in the cytoplasm, providing significant background signal when live-cell imaging. In addition, care should be taken to select monomeric fluorophores that disrupt autophagy dynamics and kinetics as little as possible; for example, the dimeric DsRed should not be used fused to autophagosome markers. We typically subclone the original Addgene constructs into vectors compatible with multi-channel live-cell imaging experiments, such as mCherry, Halo, and SNAP vectors. Halo-tagged (Promega) and SNAP-tagged (NEB) constructs result in very stable fluorophores that can be used to monitor autophagosome biogenesis and maturation over longer time scales [32]. The newer Janelia Fluors (JF) (Promega and Tocris) offer additional photostability and brightness compared to the original ligands. Halo- and SNAP-tags provide additional flexibility, as multiple ligands with different excitation/emission spectra are available; different ligands can be selected for different experiments (*see* **Note 2**).

Autophagy engulfs and degrades cargos, and this process can be directly monitored via live-cell imaging by simultaneously monitoring both autophagosome formation with an LC3B marker and a marker for the cargo being engulfed [13, 15]. Axonal autophagosomes can engulf mitochondrial fragments, which can be imaged with a construct encoding the mitochondrial targeting sequence ("mito" in DNA constructs) fused to a fluorophore, such as Mito-DsRed or Mito-SNAP. Protein aggregates can be imaged using a fluorescently tagged ubiquitin, such as mRFP-Ub, or with neurodegenerative disease-associated constructs such as mCherry-SOD1^{G93A} [13] or huntingtin constructs with normal or pathological polyglutamine repeats [15]. To specifically follow the turnover of mitochondria, two probes can be used. mt-mKeima is a construct that localizes to mitochondria and is excited at 440 nm in neutral pH, but is excited at 550 nm in low pH, allowing for the distinction between engulfed and non-engulfed mitochondria with one fluorophore [38]. Mito-QC is also a mitochondrially targeted construct with tandem mCherry- and GFP-tags. As described above, mCherry remains stable at acidic pH, but GFP is readily quenched, thus when mito-QC-labeled mitochondria are engulfed by autolysosomes, the fluorescent signal switches to red-only. This construct has been used to monitor mitophagy in mouse Purkinje cells in vivo [39].

Vital dyes have also been used to image autophagy in neurons, including Lysotracker (ThermoFisher) to image acidified autolysosomes [15], Mitotracker (ThermoFisher) to image mitochondria, and CellMask Orange (Life Technologies) to image the plasma membrane [4]. These dyes should be used with caution as they are more readily bleached over time than GFP, mCherry, Halo-tagged, or SNAP-tagged fluorophores and thus limit temporal resolution.

2.7 Data Acquisition and Analysis

Most microscopy systems are driven by either manufacturer-specific software, such as the Volocity package from PerkinElmer, or alternatively by broader platform software available commercially (MetaMorph, Molecular Devices) or open source (μManager, Open Imaging). This software will allow precise control of imaging parameters such as laser power, exposure time, offset, gain, and imaging rates. It is important to record these values on a per-image basis, and to be as consistent as possible from experiment to experiment. We strongly recommend that images be acquired with minimal processing, such as pixel averaging, and be maintained as raw, unaltered files in permanent storage, on a high-capacity external hard drive or on a secure server.

Image processing, such as cropping or contrast enhancement, should be performed to the strictest standards, as described previously [40, 41], and only on a copy of the raw data file. We use FIJI (ImageJ open-source software [42]) to analyze micrographs, including the Multiple Kymograph plugin. If necessary, images can be deconvolved using software such as Huygens (Scientific Volume Imaging). Deconvolution improves the signal-to-noise ratio, especially for dim images; removes background noise; and enhances resolution. It is critical to note that intensities cannot be measured from deconvolved images, as deconvolution algorithms do not necessarily maintain linearity of intensity measurements.

3 Methods

3.1 Imaging Parameters

Specific settings will depend on the experimental objective, as well as the expression level, brightness, and stability of the fluorophores involved. For live-cell imaging, the over-riding goal is to use the lowest level expression of the fluorophore and expose the biological sample to the smallest amount of laser light, while still obtaining a crisp high-resolution image. Decreasing the amount of irradiating light will limit the photo-induced cellular damage or toxicity. The laser power, exposure, number of z-slices, length of time lapse, and frequency of time points will all directly affect the amount of light experienced by the cell. While imaging must occur within the dynamic range of the sensitivity of the camera, increasing the sensitivity of the camera will allow less light to irradiate the sample.

For our experiments imaging autophagosome biogenesis, we typically image three channels for 10 min, acquiring a time point every 3 s in a single z-plane. Optimizing the imaging parameters and decreasing the channels has allowed us to image five z-planes every 3 s for up to 20 min in DRG axonal tips without excessive photobleaching or photodamage. For autophagosome maturation, we typically use two channels every 0.5–3 s for 3 min in a single z-plane. Exposure, camera sensitivity, and laser power will depend on the specific microscope, neuronal type, expression constructs, and transfection success. We avoid using the 405 nm (blue) laser, as the available blue fluorophores are neither as bright nor as stable as those available for other laser lines. Furthermore, the 405 nm laser line produces the highest energy light, inducing more photobleaching and phototoxicity than the other lasers. For neurons imaged in Hibernate media (BrainBits) supplemented with 2% B-27 and 2 mM GlutaMAX, we can successfully image a single dish for more than 1 h under these conditions.

Multiple channels may be acquired consecutively, or simultaneously if the microscope is equipped with a multi-channel imaging system such as Dualview. Collecting z-stacks will provide significantly more spatial resolution, essential for imaging high-volume cellular compartments such as the soma. However, the improved spatial resolution will come at the expense of temporal resolution and faster photobleaching of the fluorophores being used.

Neurons selected for imaging should be representative of the culture as a whole, demonstrating an intact morphology and no blebbing, which can be a sign of photo-induced toxicity. It is important to select neurons that are not excessively overexpressing the probes, as high overexpression could result in toxicity or disruption of the pathway. For DRG neurons in culture, all processes are axon-like, with a uniformly polarized microtubule cytoskeleton [43]. For hippocampal or cortical cultures, axons should be distinguished from dendrites prior to imaging. This can be done on the basis of morphological criteria [31]; axons are longer, thinner, and of uniform caliber along their lengths, while dendrites extend 200–300 μm from the soma with diameters that taper away from the cell body. Alternatively, axons can be labeled by expression of an AIS marker such as YFP-NavII-III [44] or labeled with anti-neurofascin [45].

To image autophagosome biogenesis and maturation in the axon, we follow axons to their distal tips to identify the region for imaging. For autophagosome maturation, we subdivide axons into subdomains. The distal axon can be defined as the region less than 100 μm from the axon terminal, the mid-axon is defined as the region greater than 100 μm from both the axon terminal and the soma, and the proximal axon is the region less than 100 μm from the soma [4, 13, 46]. We typically image the proximal and distal axon for autophagosome maturation experiments (Fig. 2). To image autophagosome biogenesis and maturation in the soma, we use the piezo stage controller to drive the z-stack in 0.15–0.3 μm increments. For colocalization studies on live samples, it is important to capture each channel at each z-slice before moving to the next image frame in the z dimension. Otherwise, the autophagosome may have moved by the time the acquisition of all the z-planes in the first channel is complete (see Note 3).

3.2 Data Analysis

Image files are opened in FIJI and can be cropped and saved as TIFFs to facilitate ease of file access. To analyze autophagosome biogenesis, we track puncta through the time-lapse video. Signal intensity of each channel can be measured at the punctum to assess kinetics of autophagosome biogenesis. To determine rates of autophagosome biogenesis, we count the number of GFP-LC3B puncta that develop or increase in size throughout the video. Colocalization as well as rates and kinetics between the different markers can also be determined. To determine the kinetics of autophagy factors during autophagosome biogenesis, we measure the intensity of each imaged channel at a punctum, following that punctum throughout the video. Average background intensity is subtracted from each measurement and then normalized to the

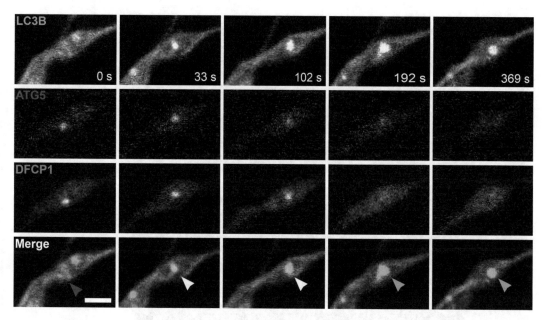

Fig. 2 Autophagosome biogenesis in the distal axon. DRG neurons were isolated from GFP-LC3B mice and transfected with DNA constructs encoding mCh-ATG5 and Halo-DFCP1. The neurons were treated with Far-red Halo ligand prior to imaging. A neuron expressing both constructs was imaged every 3 s for 10 min. The arrow in the merge panels indicates a forming autophagosome; the color of the arrow denotes the colocalization state of the tagged proteins. Scale bar, 2 μm

maximum intensity of that image sequence. If single z-planes were collected instead of z-stacks, care should be taken to ensure that puncta are not counted as they move in and out of the imaged z-plane. It is important to have high enough temporal resolution, as autophagosomes move within the axonal tip [4]; if the frequency of the time points is too low, it will be very difficult to follow puncta within the axonal tip.

To analyze autophagosome maturation, we draw a segmented line ("line scan") on top of the axon in FIJI. The segmented line should start at the same end of the axon each time (distal or proximal) to facilitate identification of anterograde and retrograde movement. If multiple channels were acquired, they need to be separated using the Split Channels function. We then use the Multiple Kymograph plugin to generate a kymograph (Fig. 3). Depending on the width of the axon, the pixel width of the line is typically set to 3 or 5 pixels (only odd-numbered pixel widths are accepted). The line can be saved as an ROI to ensure the same trace is used on all channels. ROIs can be renamed in the ROI manager and saved for future reference. Once a kymograph is generated for each channel, the transport dynamics of autophagosomes, such as velocity, direction, fusion, and acidification, can be determined by measuring the slope or intensity of the traces. It is helpful to have the original file open to play the video as kymographs are being

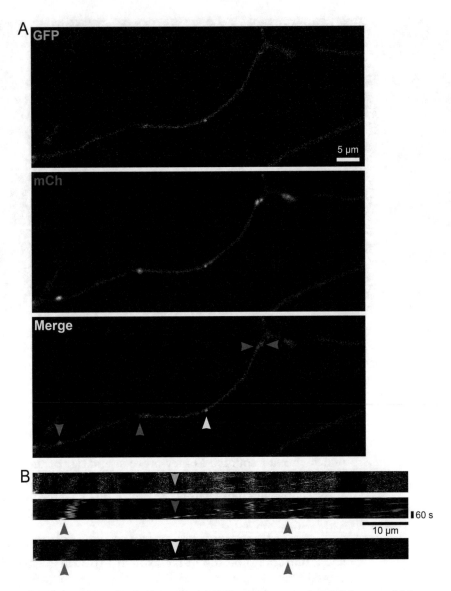

Fig. 3 Autophagosome maturation in the axon. (**a**) DRG neurons were isolated from a wild-type mouse and transfected with the tandem mCh-egfp-LC3B DNA construct. A neuron expressing the tandem construct was imaged every 3 s for 3 min. In the merge panel, the color of the arrows denotes whether or not the EGFP of the tandem construct was quenched. (**b**) Kymographs were generated using Multiple Kymograph in FIJI. The yellow arrow identifies the autophagosome containing un-quenched EGFP

analyzed to ensure that all puncta are identified and counted; this is especially true for GFP-tagged proteins that are quenched upon fusing with lysosomes, which have lower signal-to-noise ratios. We define autophagosomes as moving retrograde as puncta that have a displacement toward the soma greater than or equal to 5 μm within a 5-min imaging window. Anterogradely moving autophagosomes are defined as puncta that have a displacement toward the axonal tip greater than or equal to 5 µm within a 5-min imaging window. Autophagosomes that do not have a net displacement of 5 µm within a 5-min imaging window are classified as stationary or bidirectional [4]. It is again important to image with sufficiently high temporal resolution, as autophagosomes and lysosomes are trafficked with high velocities in the axon [13].

4   Notes

1. *Common pitfalls in neuronal culture.*

   The most challenging aspect of live-cell imaging of neurons is ensuring that the primary neuronal cultures being studied are healthy to start and remain robust throughout the imaging session. Common pitfalls in the preparation of cultures include slow or difficult dissection, and the use of media formulations that are older than 2 weeks or that are incomplete (lacking B27 is a common mistake, as it is added to media after filter sterilizing). Common pitfalls in maintaining neuronal cultures include over-handling of the cultures, which should only be checked every few days. Media should not be changed too often and never fully replaced. We typically replace no more than 50% at a time. For example, for longer term cultures we typically replace 30% of the culture media with fresh maintenance media on DIV3 and DIV7 [47]. Use of AraC effectively blocks the proliferation of glial cells, so the neurons are not overwhelmed by other cells.

2. *Common pitfalls in probe selection and expression.*

   Selection of the right probe is critical to the success of the experiment. Fluorescent tags that are not monomeric, such as DsRed, might interfere with the kinetics of the pathway. Additionally, mCherry, while a commonly used fluorescent tag, is thought to aggregate, which is a particular concern when examining a cellular system that acts to recycle protein aggregates. The relatively new mScarlet is a truly monomeric derivative of mCherry with increased brightness, quantum yield, and fluorescence lifetime [48].

   Importantly, any new construct must be tested for expression levels and successful incorporation into a bona fide autophagosome, with care taken to ensure that overexpression does not alter autophagy dynamics. Finally, it is essential that the chosen fluorophores match the excitation and emission characteristics of the microscope used for imaging.

   Practical aspects to consider are the quality of the DNA that will be transfected and the method of transfection. We prepare our DNA plasmids with miniprep or maxiprep kits, ensuring that the final DNA is endotoxin-free. Electroporation works well for DRG neurons and Lipofectamine 2000 for hippocampal neurons, but other approaches such as calcium phosphate transfection should also be considered, depending on the experiment. Lipofectamine 3000 and MessengerMAX (ThermoFisher) are other in-dish transfection kits designed for hard-to-transfect cells like neurons. Fluorophore-protein fusion constructs can be very variable in terms of expression efficiency, time required for expression, and expression levels. All of these factors must be determined empirically on a construct-by-construct basis. Neuronal cultures can be sensitive to too much transfection reagent or too much DNA, and this varies with the source of the neurons (rat or mouse) as well as the culture type (hippocampal, cortical, or DRG). Some constructs, such as Mito-DsRed or Mito-SNAP, can be transfected in low amounts, as they consistently express incredibly well in neurons.

3. *Common pitfalls in imaging.*

Commonly encountered issues during imaging include probe expression that is either too low to capture fluorescent images with sufficient signal-to-noise, or too high, leading to saturation of the signal or phototoxicity. If the imaging dish is too crowded because neurons were plated at a too high density, it can be challenging to unambiguously define the axonal and dendritic processes of a single neuron, as these processes can extend far from the soma and overlap, intersect or fasciculate on the culture dish. Lower transfection efficiencies or the mixing of labeled and unlabeled neurons in a single dish [4] can help with this. Maintaining a consistent focal plane is essential, but can readily be accomplished using a Perfect Focus (Nikon) mechanism or the equivalent if your microscope has this feature. A Perfect Focus-like system works by detecting the coverslip-media interface, so it is not capable of maintaining focus on an autophagic vesicle that moves in the z axis. Finally, care must be taken throughout the imaging session to avoid photobleaching of the signal or phototoxic damage to the neuron by optimizing the laser power, exposure, and acquisition parameters of the camera.

References

1. Hara T, Nakamura K, Matsui M, Yamamoto A, Nakahara Y, Suzuki-Migishima R, Yokoyama M, Mishima K, Saito I, Okano H et al (2006) Suppression of basal autophagy in neural cells causes neurodegenerative disease in mice. Nature 441:885–889

2. Komatsu M, Wang QJ, Holstein GR, Friedrich VL Jr, Iwata J, Kominami E, Chait BT, Tanaka K, Yue Z (2007) Essential role for autophagy protein Atg7 in the maintenance of axonal homeostasis and the prevention of axonal degeneration. Proc Natl Acad Sci U S A 104:14489–14494

3. Wu Y, Whiteus C, Xu CS, Hayworth KJ, Weinberg RJ, Hess HF, De Camilli P (2017) Contacts between the endoplasmic reticulum and other membranes in neurons. Proc Natl Acad Sci U S A 114:E4859–E4867

4. Maday S, Holzbaur EL (2014) Autophagosome biogenesis in primary neurons follows an ordered and spatially regulated pathway. Dev Cell 30:71–85

5. Gowrishankar S, Yuan P, Wu Y, Schrag M, Paradise S, Grutzendler J, De Camilli P, Ferguson SM (2015) Massive accumulation of luminal protease-deficient axonal lysosomes at Alzheimer's disease amyloid plaques. Proc Natl Acad Sci U S A 112:E3699–E3708

6. Nirschl JJ, Ghiretti AE, Holzbaur ELF (2017) The impact of cytoskeletal organization on the local regulation of neuronal transport. Nat Rev Neurosci 18:585–597

7. Huang CY, Rasband MN (2018) Axon initial segments: structure, function, and disease. Ann N Y Acad Sci 1420:46–61

8. Leterrier C (2018) The axon initial segment: an updated viewpoint. J Neurosci 38:2135–2145

9. Boecker CA, Holzbaur EL (2019) Vesicular degradation pathways in neurons: at the crossroads of autophagy and endo-lysosomal degradation. Curr Opin Neurobiol 57:94–101

10. Mizushima N, Yamamoto A, Matsui M, Yoshimori T, Ohsumi Y (2004) In vivo analysis of autophagy in response to nutrient starvation using transgenic mice expressing a fluorescent autophagosome marker. Mol Biol Cell 15:1101–1111

11. Hollenbeck PJ (1993) Products of endocytosis and autophagy are retrieved from axons by regulated retrograde organelle transport. J Cell Biol 121:305–315

12. Lee S, Sato Y, Nixon RA (2011) Lysosomal proteolysis inhibition selectively disrupts axonal transport of degradative organelles and causes an Alzheimer's-like axonal dystrophy. J Neurosci 31:7817–7830

13. Maday S, Wallace KE, Holzbaur EL (2012) Autophagosomes initiate distally and mature during transport toward the cell soma in primary neurons. J Cell Biol 196:407–417

14. Cheng XT, Zhou B, Lin MY, Cai Q, Sheng ZH (2015) Axonal autophagosomes recruit dynein for retrograde transport through fusion with late endosomes. J Cell Biol 209:377–386

15. Wong YC, Holzbaur EL (2014) The regulation of autophagosome dynamics by huntingtin and HAP1 is disrupted by expression of mutant huntingtin, leading to defective cargo degradation. J Neurosci 34:1293–1305

16. Fu M-M, Nirschl JJ, Holzbaur ELF (2014) LC3 binding to the scaffolding protein JIP1 regulates processive dynein-driven transport of autophagosomes. Dev Cell 29:577–590

17. Stavoe AK, Hill SE, Hall DH, Colon-Ramos DA (2016) KIF1A/UNC-104 transports ATG-9 to regulate neurodevelopment and autophagy at synapses. Dev Cell 38:171–185

18. Neisch AL, Neufeld TP, Hays TS (2017) A STRIPAK complex mediates axonal transport of autophagosomes and dense core vesicles through PP2A regulation. J Cell Biol 216:441–461

19. Jin EJ, Kiral FR, Ozel MN, Burchardt LS, Osterland M, Epstein D, Wolfenberg H, Prohaska S, Hiesinger PR (2018) Live observation of two parallel membrane degradation pathways at axon terminals. Curr Biol 28:1027–1038.e1024

20. Koyama-Honda I, Itakura E, Fujiwara TK, Mizushima N (2013) Temporal analysis of recruitment of mammalian ATG proteins to the autophagosome formation site. Autophagy 9:1491–1499

21. Itakura E, Mizushima N (2010) Characterization of autophagosome formation site by a hierarchical analysis of mammalian Atg proteins. Autophagy 6:764–776

22. Maday S, Holzbaur EL (2016) Compartment-specific regulation of autophagy in primary neurons. J Neurosci 36:5933–5945

23. Tang G, Gudsnuk K, Kuo SH, Cotrina ML, Rosoklija G, Sosunov A, Sonders MS, Kanter E, Castagna C, Yamamoto A et al (2014) Loss of mTOR-dependent macroautophagy causes autistic-like synaptic pruning deficits. Neuron 83:1131–1143

24. Devireddy S, Liu A, Lampe T, Hollenbeck PJ (2015) The organization of mitochondrial quality control and life cycle in the nervous system in vivo in the absence of PINK1. J Neurosci 35:9391–9401

25. Sung H, Tandarich LC, Nguyen K, Hollenbeck PJ (2016) Compartmentalized regulation of Parkin-mediated mitochondrial quality control in the Drosophila nervous system in vivo. J Neurosci 36:7375–7391

26. Ashrafi G, Schlehe JS, LaVoie MJ, Schwarz TL (2014) Mitophagy of damaged mitochondria occurs locally in distal neuronal axons and requires PINK1 and Parkin. J Cell Biol 206:655–670

27. Evans CS, Holzbaur ELF (2019) Autophagy and mitophagy in ALS. Neurobiol Dis 122:35–40

28. Fernandopulle MS, Prestil R, Grunseich C, Wang C, Gan L, Ward ME (2018) Transcription factor-mediated differentiation of human iPSCs into neurons. Curr Protoc Cell Biol 79:e51

29. Hsieh CH, Shaltouki A, Gonzalez AE, Bettencourt da Cruz A, Burbulla LF, St Lawrence E, Schule B, Krainc D, Palmer TD, Wang X (2016) Functional impairment in Miro degradation and mitophagy is a shared feature in familial and sporadic Parkinson's disease. Cell Stem Cell 19:709–724

30. Hill SE, Kauffman KJ, Krout M, Richmond JE, Melia TJ, Colon-Ramos DA (2019) Maturation and clearance of autophagosomes in neurons depends on a specific cysteine protease isoform, ATG-4.2. Dev Cell 49(2):251–266. e8

31. Kaech S, Banker G (2006) Culturing hippocampal neurons. Nat Protoc 1:2406–2415

32. Stavoe AK, Holzbaur EL (2018) Expression of WIPI2B counteracts age-related decline in autophagosome biogenesis in neurons. Elife 8:e44219

33. Pacifici M, Peruzzi F (2012) Isolation and culture of rat embryonic neural cells: a quick protocol. J Vis Exp 63:e3965

34. Perlson E, Jeong G-B, Ross JL, Dixit R, Wallace KE, Kalb RG, Holzbaur ELF (2009). A switch in retrograde signaling from survival to stress in rapid-onset neurodegeneration. J Neurosci 29(31):9903–9917. PMCID: PMC3095444

35. Jiang M, Chen G (2006) High Ca^{2+}-phosphate transfection efficiency in low-density neuronal cultures. Nat Protoc 1:695–700

36. Pankiv S, Clausen TH, Lamark T, Brech A, Bruun JA, Outzen H, Overvatn A, Bjorkoy G, Johansen T (2007) p62/SQSTM1 binds directly to Atg8/LC3 to facilitate degradation of ubiquitinated protein aggregates by autophagy. J Biol Chem 282:24131–24145

37. Li J, Chen Z, Stang MT, Gao W (2017) Transiently expressed ATG16L1 inhibits autophagosome biogenesis and aberrantly targets RAB11-positive recycling endosomes. Autophagy 13:345–358

38. Katayama H, Kogure T, Mizushima N, Yoshimori T, Miyawaki A (2011) A sensitive and quantitative technique for detecting autophagic events based on lysosomal delivery. Chem Biol 18:1042–1052

39. McWilliams TG, Prescott AR, Allen GF, Tamjar J, Munson MJ, Thomson C, Muqit MM, Ganley IG (2016) mito-QC illuminates mitophagy and mitochondrial architecture in vivo. J Cell Biol 214:333–345

40. Rossner M, Yamada KM (2004) What's in a picture? The temptation of image manipulation. J Cell Biol 166:11–15

41. North AJ (2006) Seeing is believing? A beginners' guide to practical pitfalls in image acquisition. J Cell Biol 172(1):9–18

42. Schindelin J, Arganda-Carreras I, Frise E, Kaynig V, Longair M, Pietzsch T, Preibisch S, Rueden C, Saalfeld S, Schmid B et al (2012) Fiji: an open-source platform for biological-image analysis. Nat Methods 9:676–682

43. Moughamian AJ, Osborn GE, Lazarus JE, Maday S, Holzbaur EL (2013) Ordered recruitment of dynactin to the microtubule plus-end is required for efficient initiation of retrograde axonal transport. J Neurosci 33:13190–13203

44. Grubb MS, Burrone J (2010) Activity-dependent relocation of the axon initial segment fine-tunes neuronal excitability. Nature 465:1070–1074

45. Aiken J, Moore JK, Bates EA (2018) TUBA1A mutations identified in lissencephaly patients dominantly disrupt neuronal migration and impair dynein activity. Hum Mol Genet 28 (8):1227–1243

46. Olenick MA, Dominguez R, Holzbaur ELF (2019) Dynein activator Hook1 is required for trafficking of BDNF-signaling endosomes in neurons. J Cell Biol 218:220–233

47. Guedes-Dias P, Nirschl JJ, Abreu N, Tokito MK, Janke C, Magiera MM, Holzbaur ELF (2019) Kinesin-3 responds to local microtubule dynamics to target synaptic cargo delivery to the presynapse. Curr Biol 29:268–282.e268

48. Bindels DS, Haarbosch L, van Weeren L, Postma M, Wiese KE, Mastop M, Aumonier S, Gotthard G, Royant A, Hink MA et al (2017) mScarlet: a bright monomeric red fluorescent protein for cellular imaging. Nat Methods 14:53–56

Chapter 4

Monitoring Autophagic Activity In Vitro and In Vivo Using the GFP-LC3-RFP-LC3ΔG Probe

Tomoya Eguchi, Hideaki Morishita, and Noboru Mizushima

Abstract

Autophagy is a major degradative pathway influencing various biological processes and diseases. A method for measuring autophagic activity is essential to elucidate the roles and mechanisms of autophagy. Although several relevant methods have been reported, the quantitative monitoring of autophagic activity in cells and in animals remains challenging. This chapter provides methods involving the recently developed autophagic flux probe GFP-LC3-RFP-LC3ΔG. This probe is cleaved to produce GFP-LC3 and RFP-LC3ΔG, the former of which is degraded by autophagy, and the latter of which is maintained in the cytoplasm to serve as an internal control. Autophagic activity can be simply and quantitatively estimated by determining the ratio of GFP and RFP signal intensities. The probe is appropriate for cell cultures as well as in vivo analyses. In this chapter, we describe the utility of this probe for measuring autophagic activity in cultured cells by flow cytometry and for analyzing autophagic activity in zebrafish tissues by confocal microscopy.

Keywords Autophagy, Lysosome, LC3, Flow cytometry, Zebrafish

1 Introduction: Macroautophagy

Macroautophagy (hereafter called autophagy) refers to a major intracellular degradative pathway, in which a portion of the cytoplasm and organelles are sequestered by the isolation membrane and transported into lysosomes. Autophagy is involved in various physiological processes, including the adaptation to starvation and the quality control of cells and tissues, as well as the pathology of various diseases [1, 2]. An analysis of autophagic activity is essential to understand the physiological and pathological roles of autophagy and to elucidate the underlying molecular mechanisms. Thus, several methods for monitoring autophagic activity (also called autophagic flux) have been reported [3, 4].

The simplest method for estimating autophagic activity may involve counting the autophagosomes carrying LC3, an autophagosomal marker [5], by fluorescence microscopy. Alternatively, the abundance of LC3 conjugated to phosphatidylethanolamine (also

Ben Loos and Esther Wong (eds.), *Imaging and Quantifying Neuronal Autophagy*, Neuromethods, vol. 171,
https://doi.org/10.1007/978-1-0716-1589-8_4, © Springer Science+Business Media, LLC, part of Springer Nature 2022

known as LC3-II), which is localized to the autophagosomal membranes, can be biochemically quantified. However, the accumulation of autophagosomes or LC3-II indicates not only that autophagy has been activated, but also that the degradation of autophagosomes is impaired (e.g., lysosomal dysfunction). Therefore, the data should be carefully interpreted to distinguish whether autophagy is activated or inhibited.

A conventional way to distinguish between the activation of autophagy and the impairment of autophagosome degradation involves comparing the amount of autophagosomes or LC3-II in cells with or without lysosomal inhibitor treatment. The differences in the amounts between these conditions reflect the amount of degraded autophagosomes when lysosomes are inhibited. Although this method is widely used, there are some limitations. First, lysosomal inhibitors suppress mTORC1 (mechanistic target of rapamycin complex 1), which may then induce autophagy [6, 7]. Additionally, lysosomotropic reagents such as chloroquine induce the conversion of LC3-I (unconjugated form of LC3) to LC3-II independently of autophagy [8]. Another problem with this method is that monitoring autophagic activity in whole organisms is difficult because complete inhibition of lysosomal activity in vivo is impractical.

Fluorescent protein-based autophagic flux probes have recently been reported. The first such probe was RFP-GFP-LC3 (tandem fluorescent-tagged LC3; tfLC3) [9]. The tfLC3 is converted to a lipidated form on the autophagosomal membrane and transported to lysosomes, where the fluorescence of GFP is readily quenched. However, RFP is more resistant to lysosomal degradation. Therefore, RFP single-positive punctate structures indicate autolysosomes, whereas GFP-RFP double-positive structures indicate autophagosomes prior to lysosomal degradation. The accumulation of RFP single-positive structures may imply autophagy has been activated. This probe is, therefore, useful to determine the maturation stage of individual autophagic structures.

The second probe described was Keima [10], which is an acid-stable fluorescent protein whose excitation spectrum changes according to the pH of the surrounding environment. The excitation spectrum has a peak at 440 nm under neutral conditions and 568 nm under acidic conditions. The amount of Keima imported into lysosomes can be estimated based on the ratio of fluorescent signal excited by these wavelengths. Although Keima enables the quantitative monitoring of autophagic activity, a conventional fluorescence microscope is usually not equipped with the lasers and filters needed to measure Keima fluorescence.

The most recently described probe is GFP-LC3-RFP-LC3ΔG [11, 12], which has been used in cell biological studies [13–15] as well as for genome-wide CRISPR screening [16]. In this chapter, we describe the principle underlying this novel probe and its utility for monitoring autophagic activity in cultured cells and in zebrafish spinal cord.

2   Materials

2.1   Cell Culture

Wild-type (WT) and *Atg5* knockout (KO) mouse embryonic fibroblast (MEF) cells were prepared as previously described in [17]. Additionally, HEK293T cells were obtained from the American Type Culture Collection. Cells were plated on tissue culture dishes (Falcon) and cultured in Dulbecco's modified Eagle medium (DMEM; Sigma-Aldrich) supplemented with 10% (v/v) fetal bovine serum (FBS; Sigma-Aldrich) and 2 mM L-glutamine (Thermo Fisher Scientific) at 37 °C in a 5% CO_2 atmosphere. To activate starvation-induced autophagy, cells were cultured in DMEM without amino acids and FBS (Wako).

2.2   Establishment of Cells Stably Expressing the Probe Using the Retrovirus Vector

The pMRX-IP-GFP-LC3-RFP-LC3ΔG plasmid, which was constructed as previously described in [12], is available from Addgene (plasmid 84572) and the Riken BioResource Center (RDB14600). The pMRX-IP vector was provided by Dr. Shoji Yamaoka [18], whereas the pCG-VSV-G and pCG-gag-pol vectors were provided by Dr. Teruhito Yasui. To construct a retrovirus vector, HEK293T cells were transfected with pMRX-IP-GFP-LC3-RFP-LC3ΔG, pCG-VSV-G, and pCG-gag-pol using Lipofectamine 2000 (Thermo Fisher Scientific). At 1 day after transfection, the medium was changed and the cells were cultured for another 2 days. The retrovirus-containing medium was then collected and filtered through a 0.45-μm pore PVDF membrane (Millipore) to remove cell debris. The filtered medium can be stored at 4 °C for at least 3 days. For long-term storage, the medium can be kept at −80 °C; however, one freeze-thaw cycle may decrease the virus titer by half. The retrovirus should be handled under the appropriate environmental conditions.

To obtain infected cells, MEF cells were cultured for 1 day in medium supplemented with 50% (v/v) retrovirus-containing medium and 8 μg/mL polybrene (Sigma-Aldrich). The medium was changed to fresh DMEM containing 2–3 μg/mL puromycin (Sigma-Aldrich). Single clones from infected cells were isolated by limiting dilution. The production of the full-length GFP-LC3-RFP-LC3ΔG probe was confirmed by genomic PCR and western blot.

2.3   Genomic PCR

Genomic DNA was purified using a Blood & Cell Culture DNA Midi Kit (Qiagen). The purified genomic DNA was mixed with primers and PrimeSTAR Max Premix (Takara) for a PCR assay. The following primers were used to amplify the GFP-LC3-RFP-LC3ΔG fragment: 5'- CGCCGCCGGGATCACTCTCG -3' and 5'- CCACCACACTGGGATCCTTA -3'. The PCR product sizes were determined by agarose gel electrophoresis.

2.4 Flow Cytometry

The EC800 flow cytometer (Sony) equipped with 488-nm and 561-nm lasers was used to quantify the probe fluorescence intensity in the cultured cells. Before cells expressing the probe were analyzed, the fluorescence compensation process was completed by analyzing cells expressing GFP or RFP alone. At least 10,000 events were acquired for each sample.

2.5 Zebrafish Preparation

Wild-type zebrafish were purchased from Riken and maintained at 28 °C under a 14-h light/10-h dark photoperiod with the Labreed system (Iwaki). Additionally, *fip200/rb1cc1* KO zebrafish were produced using the CRISPR/Cas9 system [12]. Briefly, single-cell stage zebrafish eggs were injected with Alt-R Cas9 nuclease (Integrated DNA Technologies) as well as Alt-R CRISPR-Cas9 tracrRNA (Integrated DNA Technologies) and crRNA (Integrated DNA Technologies) to target the following sequence in exon 4 of zebrafish *fip200*: 5′- GATGAACACCTTCAGCACCA -3′. The genotype was confirmed by a heteroduplex mobility assay [19] using genomic DNA, PrimeSTAR Max Premix, and primers 5′-A GCTTTGCTCGTTCTGTGAGCG -3′ and 5′- CTGATAG GAAAGGGTGCAGTCG -3′. The amplified fragments were analyzed by 10% polyacrylamide gel electrophoresis.

2.6 In Vitro Transcription

The pcDNA3-GFP-LC3-RFP-LC3ΔG plasmid, which was constructed as previously described in [12], is available from Addgene (plasmid 168997). The GFP-LC3-RFP-LC3ΔG sequence was cloned into pcDNA3 so that it was sandwiched between an upstream SP6 promoter and a downstream polyadenylation signal. The plasmid was linearized with XhoI, after which the capped mRNA with a poly-A tail was synthesized using the mMESSAGE mMACHINE Ultra Kit (Thermo Fisher Scientific). The synthesized mRNA was purified using an RNeasy Mini Prep Kit (Qiagen) and suspended in RNase-free water. The purified mRNA was stored at −80 °C.

2.7 Injection of mRNA into Zebrafish Eggs

On the day before the injection, male and female zebrafish were placed in the same tank but were separated by a barrier. The next morning, the barrier was removed and fresh fertilized eggs were obtained. The mRNA stored at −80 °C was diluted in RNase-free water for a final concentration of 50 ng/μL, after which it was maintained on ice until the injection. The mRNA solution was added to a FemtoTip II glass capillary (Eppendorf), and a FemtoJet (Eppendorf) was used to inject the mRNA into eggs at the single-cell stage. The eggs were incubated in water supplemented with methylene blue as an antiseptic. Additionally, 0.1875% (w/v) 1-phenyl 2-thiourea (PTU) (Sigma-Aldrich) was dissolved in distilled water to prepare a 50× stock solution, which was stored at 4 °C in the dark.

2.8 Confocal Microscopy

Fluorescence images of zebrafish anesthetized with 0.03% tricaine (Sigma-Aldrich, A5040) were acquired using a FV1000 IX81 confocal microscope (Olympus) equipped with a UPLSAPO30XS objective lens (Olympus). The images were then analyzed using ImageJ and MetaMorph (Molecular Devices).

3 Methods

3.1 Principle Underlying the Novel Probe GFP-LC3-RFP-LC3Δ G

The GFP-LC3-RFP-LC3ΔG probe was designed to quantitatively assess the autophagic activity in cells and in whole organisms, including mice and zebrafish. The probe is cleaved by the endogenous proteinase ATG4 to produce an equimolar amount of GFP-LC3 and RFP-LC3ΔG (Fig. 1a). The GFP-LC3 is conjugated to phosphatidylethanolamine on the autophagosomal membrane and degraded in autolysosomes. In contrast, RFP-LC3ΔG, whose C-terminal glycine is deleted, remains in the cytoplasm because the glycine is essential for lipidation. After autophagy is induced, the

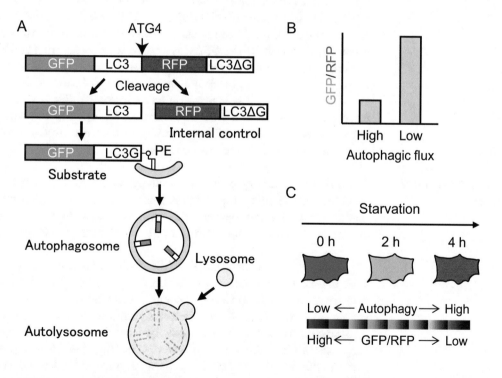

Fig. 1 Method for measuring autophagic activity using the GFP-LC3-RFP-LC3ΔG probe. (From Morishita et al. [11]). (**a**) GFP-LC3-RFP-LC3ΔG is cleaved by the endogenous protease ATG4 into GFP-LC3 (an autophagic substrate) and RFP-LC3ΔG (an internal control). (**b**) Autophagic activity is estimated by calculating the ratio of GFP and RFP signal. The GFP:RFP ratio decreases when the autophagic flux is high, whereas it increases when the autophagic flux is low. (**c**) Schematic representation of the starvation-induced autophagy in cells expressing GFP-LC3-RFP-LC3ΔG. The blue signal represents a high GFP:RFP ratio (low autophagic activity), while the red signal represents a low GFP:RFP ratio (high autophagic activity)

GFP signal decreases, whereas the RFP signal remains stable and serves as an internal control. The ratio of GFP and RFP signal intensities negatively correlates with autophagic activity. Specifically, a low GFP:RFP ratio corresponds to high autophagic activity, whereas a high GFP:RFP ratio indicates low autophagic activity (Fig. 1b, c).

3.2 Analysis of Autophagic Activity in Culture Cells

3.2.1 Preparation of Probe-Expressing Cells

To ensure constant and moderate probe expression levels, cell lines stably expressing the probe should be established with a retroviral vector or via other methods. The GFP-LC3-RFP-LC3ΔG sequence may undergo homologous recombination during viral production or may be integrated into the genome of the infected cells, resulting in the expression of GFP-LC3ΔG, which lacks the internal LC3-RFP. Therefore, single clones expressing the full-length GFP-LC3-RFP-LC3ΔG sequence should be isolated. The expression of the probe can be confirmed via genomic PCR and immunoblotting. The two primers described in the Subheading 2.3 can be used to amplify a 1533-bp fragment from the full-length GFP-LC3-RFP-LC3ΔG and a 480-bp fragment from GFP-LC3ΔG. Instead of GFP-LC3-RFP-LC3ΔG, GFP-LC3-RFP (without LC3ΔG) can be used to prevent homologous recombination, although RFP-LC3ΔG is a more acculate internal control (*see* **Note 1**). The probe expression levels should also be considered. High probe expression levels may mask the degradation of GFP by autophagy, whereas low expression levels make it difficult to accurately measure GFP and RFP fluorescence. Comparing the autophagic activity between different cell groups with similar probe expression levels is also recommended.

3.2.2 Analysis by Flow Cytometry

The probe-expressing cells can be analyzed by several methods, including those involving a fluorescence microscope, microplate reader, or a flow cytometer [12]. Among these methods, flow cytometry enables high-throughput and highly quantitative analyses.

Here, wild-type or *Atg5* KO MEF cells expressing the probe were exposed to starvation conditions for 4 h. The optimal starvation treatment duration varies depending on the cell line. Longer starvation periods result in a better signal-to-noise ratio but are associated with increased cytotoxicity. Trypsin-treated cells were suspended in ice-cold PBS. Cells were centrifuged at $1500 \times g$ for 2 min, after which the pelleted cells were resuspended in ice-cold PBS lacking Ca^{2+} to prevent cell aggregation. Cells were passed through a cell strainer to remove cell aggregates and then transferred to a collection tube. The harvested cells should be analyzed as soon as possible. The GFP:RFP signal ratio for this probe in MEF cells is relatively stable for at least 3 h when cells are kept on ice. The GFP and RFP fluorescence in cells was analyzed using a cell analyzer. For example, GFP was excited by a 488-nm laser and

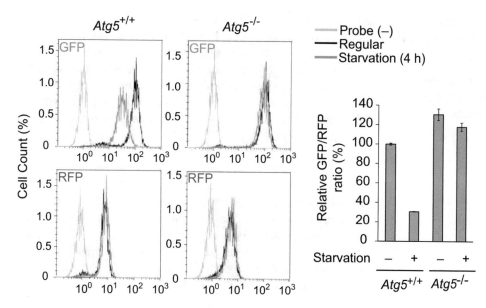

Fig. 2 Examples of histograms of fluorescence intensity versus the cell count and the GFP:RFP ratio (as a percentage of the mean of cells cultured in regular medium). (From Kaizuka et al. [12])

fluorescence was detected at 500–550-nm wavelengths, whereas RFP was excited by a 561-nm laser and fluorescence was monitored at 565–605-nm wavelengths. An appropriate fluorescence filter as well as fluorescence compensation are essential for quantifying autophagic activity with this probe.

A representative result is shown in Fig. 2. In WT cells expressing the probe, decreased GFP signal was detected in response to starvation, whereas RFP signal was unchanged. This decrease in the GFP signal was not observed in the *Atg5* KO MEF cells, indicating that the signal change reflects autophagic activity. The GFP:RFP ratio was calculated (the mean of the GFP signal was divided by the mean of the RFP signal) as an indicator of autophagic activity. The ratio negatively correlates with autophagic activity. The ratio data were normalized against the ratio in the WT cells under normal growth conditions (Fig. 2). The GFP:RFP ratio decreased in response to starvation in the WT cells, but was essentially unaffected in the *Atg5* KO cells. However, the GFP:RFP ratio may decrease by 10–30% under starvation conditions even in autophagy-deficient cells, although the underlying mechanism is currently unknown. When using the probe to analyze the effects of specific treatments on autophagy, autophagy-deficient cells should be used as a negative control.

3.3 Analysis of Autophagic Activity in Zebrafish

3.3.1 Preparation of Zebrafish Expressing the GFP-LC3-RFP-LC3ΔG Probe

Because GFP-LC3-RFP-LC3ΔG is a simple and accurate fluorescence-based probe, it can be used to monitor the autophagic activity of a whole organism. Zebrafish is a widely used animal model, whose genome can be easily manipulated. Additionally, during development, zebrafish have a translucent body, which enables the live imaging of the probe throughout the whole body. The method for measuring the autophagic activity in this useful animal model is as follows.

First, zebrafish expressing the probe were prepared. Single-cell stage-fertilized zebrafish eggs were injected with GFP-LC3-RFP-LC3ΔG mRNA synthesized by in vitro transcription. The mRNA concentration for the injection was adjusted to 50 ng/μL to ensure low expression levels. For a whole-body analysis of zebrafish, the probe expression levels should be low. The injected eggs were placed in water containing a preservative and incubated at 28 °C. As a negative control, eggs that were not injected with GFP-LC3-RFP-LC3ΔG mRNA were prepared.

3.3.2 Analysis of Zebrafish

Zebrafish expressing GFP-LC3-RFP-LC3ΔG were analyzed using confocal microscopy. At 24 hours postfertilization, the probe signal can be easily detected because of low autofluorescence (*see* **Note 2**). The eggshell can be peeld off and removed with tweezers. Fish were placed on a glass bottom dish containing water supplemented with 0.03% (w/v) tricaine (an anesthetic). Within minutes, the zebrafish were immobilized. The zebrafish body was gently manipulated with a soft stick (e.g., a gel-loading tip) so that the lateral side of the body was positioned on the glass bottom dish. To obtain images, the zebrafish should be in a horizontal position. The glass bottom dish was set on the stage of a fluorescence microscope, and fluorescence and differential interference contrast (DIC) images were captured. The GFP was excited by a 473-nm laser and the fluorescence at 430–455 nm was detected. In contrast, the RFP was excited by a 559-nm laser and the fluorescence at 575–675 nm was recorded. GFP and RFP signal was sequentially acquired. The laser power should be as low as possible to avoid photo damage and bleaching. Moreover, the signal should not be saturated.

In addition to the images of probe-expressing zebrafish, images of zebrafish lacking the probe should be obtained as negative controls. If the background signal is relatively high, it should be subtracted from the GFP and RFP signal. In some tissues, such as the iris and the pigment epithelium of the eyes, autofluorescent signal is very high. Thus, these tissues should be excluded from the analysis.

The acquired data were analyzed using MetaMorph imaging software. To calculate the GFP:RFP ratio, the intensity of the GFP signal in each pixel was divided by the intensity of the RFP signal in the corresponding pixel using the "Arithmetic" function of Meta-Morph. The resulting GFP:RFP ratio for each pixel was visualized in a grayscale image, which was subsequently converted to an RGB

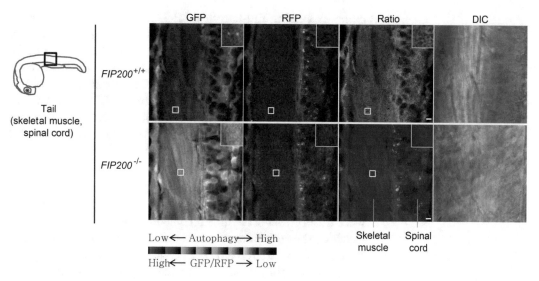

Fig. 3 Representative images of GFP and RFP fluorescence signal, GFP:RFP ratio, and DIC in the tail (skeletal muscle and spinal cord) of *fip200*+/+ and *fip200*−/− zebrafish embryos injected with GFP-LC3-RFP-LC3ΔG mRNA at 1 day postfertilization. (From Kaizuka et al. [12]). Images of the indicated regions are presented in the insets. Scale bar: 10 μm

pseudo-color image (Fig. 3). To calculate the GFP:RFP ratio in each tissue, the mean intensity of the GFP and RFP signal per cell area was obtained using Image J. The mean intensity of the GFP signal was divided by the mean intensity of the RFP signal. Alternatively, the GFP:RFP ratio for each pixel can be determined using MetaMorph.

Representative data are shown in Fig. 3. In this study, WT and *fip200* KO zebrafish expressing GFP-LC3-RFP-LC3ΔG were compared at 1 day postfertilization. In the WT zebrafish, the GFP:RFP ratio was lower for the lens and higher for the spinal cord relative to the ratio for the retina and skeletal muscles. These differences were not observed for the *fip200* KO zebrafish, in which the ratio was relatively high, indicating that these ratio differences reflect the variability of the autophagic activity in each tissue.

4   Notes

1. *GFP-LC3-RFP-LC3ΔG and GFP-LC3-RFP.*

   We previously determined that GFP-LC3-RFP (without LC3ΔG) and GFP-LC3-RFP-LC3ΔG function similarly [12]. As mentioned above, the GFP-LC3-RFP-LC3ΔG sequence may undergo homologous recombination and lose its LC3-RFP sequence. We estimated that after a viral infection and drug selection, the proportion of cells expressing GFP-LC3ΔG can reach 50%. Therefore, isolating a single clone expressing full-length GFP-LC3-RFP-LC3ΔG is essential. On the other hand, the GFP-LC3-RFP sequence does not undergo homologous recombination and, therefore, can be used immediately after cells are selected by antibiotic treatment. Although RFP-LC3ΔG is an accurate internal control that may mimic autophagy-independent but LC3-specific degradation, our analysis indicates that both probes function similarly. GFP-LC3-RFP is available from Addgene (plasmid 84573) and the Riken BioResource Center (RDB14601).

2. *Analysis of adult zebrafish.*

Zebrafish tissue exhibits strong autofluorescence, which disrupts analysis of the probe. As mentioned above, zebrafish were observed at 1 day postfertilization, which is a time point in which the autofluorescence is observed in only a few tissues. If zebrafish are analyzed at later time points (e.g., 3 days postfertilization), they should be treated with PTU to inhibit melanin production [20]. Specifically, at 1 day after mRNA injection, the eggs should be dipped in water supplemented with 0.00375% (w/v) PTU. Unfortunately, adult zebrafish cannot be analyzed because of excessive autofluorescence.

References

1. Choi AMK, Ryter SW, Levine B (2013) Autophagy in human health and disease. N Engl J Med 368(19):1845–1846

2. Mizushima N, Komatsu M (2011) Autophagy: renovation of cells and tissues. Cell 147(4):728–741

3. Klionsky DJ et al (2016) Guidelines for the use and interpretation of assays for monitoring autophagy (3rd edition). Autophagy 12(1):1–222

4. Mizushima N, Yoshimori T, Levine B (2010) Methods in mammalian autophagy research. Cell 140(3):313–326

5. Kabeya Y et al (2000) LC3, a mammalian homologue of yeast Apg8p, is localized in autophagosome membranes after processing. EMBO J 19(21):5720–5728

6. Juhász G (2012) Interpretation of bafilomycin, pH neutralizing or protease inhibitor treatments in autophagic flux experiments. Autophagy 8(12):2–1

7. Zoncu R et al (2011) mTORC1 senses lysosomal amino acids through an inside-out mechanism that requires the vacuolar H(+)-ATPase. Science 334(6056):678–683

8. Florey O, Gammoh N, Kim SE, Jiang X, Overholtzer M (2015) V-ATPase and osmotic imbalances activate endolysosomal LC3 lipidation. Autophagy 11(1):88–99

9. Kimura S, Noda T, Yoshimori T (2007) Dissection of the autophagosome maturation process by a novel reporter protein, tandem fluorescent-tagged LC3. Autophagy 3(5):452–460

10. Katayama H, Kogure T, Mizushima N, Yoshimori T, Miyawaki A (2011) A sensitive and quantitative technique for detecting autophagic events based on lysosomal delivery. Chem Biol 18(8):1042–1052

11. Morishita H, Kaizuka T, Hama Y, Mizushima N (2017) A new probe to measure autophagic flux in vitro and in vivo. Autophagy 13(4):757–758

12. Kaizuka T et al (2016) An autophagic flux probe that releases an internal control. Mol Cell 64(4):835–849

13. Carrascoso I, Alcalde J, Sánchez-Jiménez C, González-Sánchez P, Izquierdo JM (2017) T-cell intracellular antigens and Hu antigen R antagonistically modulate mitochondrial activity and dynamics by regulating optic atrophy 1 gene expression. Mol Cell Biol 37(17):17

14. Nguyen TB et al (2017) DGAT1-dependent lipid droplet biogenesis protects mitochondrial function during starvation-induced autophagy. Dev Cell 42(1):9–2100000

15. Tamura N et al (2017) Differential requirement for ATG2A domains for localization to

autophagic membranes and lipid droplets. FEBS Lett 591(23):3819–3830

16. Morita K et al (2018) Genome-wide CRISPR screen identifies TMEM41B as a gene required for autophagosome formation. J Cell Biol 217 (11):3817–3828

17. Kuma A et al (2004) The role of autophagy during the early neonatal starvation period. Nature 432(7020):1032–1036

18. Saitoh T, Nakano H, Yamamoto N, Yamaoka S (2002) Lymphotoxin-beta receptor mediates NEMO-independent NF-kappaB activation. FEBS Lett 532(1–2):45–51

19. Ota S, Hisano Y, Ikawa Y, Kawahara A (2014) Multiple genome modifications by the CRISPR/Cas9 system in zebrafish. Genes Cells 19(7):555–564

20. Karlsson J, von Hofsten J, Olsson P-E (2001) Generating transparent zebrafish: a refined method to improve detection of gene expression during embryonic development. Mar Biotechnol 3(6):522–527

Chapter 5

Measuring Autophagic Flux in Neurons by Optical Pulse Labeling

Nicholas A. Castello and Steven Finkbeiner

Abstract

Numerous assays have been developed to monitor different stages or components of autophagy. However, relatively few methods exist for quantifying overall flow through the autophagy process, known as autophagic flux. Most commonly, autophagic flux is measured using a metabolic pulse-chase (MPC) assay, which labels a subset of proteins ("pulse") and then measures levels of these proteins at regular intervals ("chase") to estimate the mean half-life of proteins cleared by autophagy. In this chapter, we describe an alternative, imaging-based method for measuring autophagic flux, called optical pulse labeling (OPL). OPL offers a number of advantages over the MPC flux assay, namely (1) the ability to calculate protein half-lives in single cells, which allows analysis of subpopulations within a heterogeneous sample, and (2) the ability to multiplex with other assays and create extremely rich datasets. This chapter provides an overview of the advantages and limitations of OPL, and a detailed protocol for carrying out OPL in primary rodent neurons.

Key words Autophagic flux, Photoconvertible fluorescent reporter, Optical pulse labeling, Time-lapse microscope, Half-life, Primary neurons

1 Introduction

Macroautophagy, herein referred to as autophagy, is a multistage process involving an interaction between several cellular compartments and numerous protein complexes. Many assays have been developed that probe individual components or stages of autophagy, but most of these assays on their own are unable to assess overall autophagic function [1, 2]. For example, measuring lipidation levels of the microtubule-associated protein light chain 3 (LC3) by western blot analysis is commonly used to detect a change in autophagy function. However, this assay alone cannot determine whether a detected change is related to induction or impairment of autophagy [2]. Thus, a western blot for LC3 lipidation without the presence of lysosomal inhibitors cannot distinguish between, for example, enhanced autophagosome formation

Ben Loos and Esther Wong (eds.), *Imaging and Quantifying Neuronal Autophagy*, Neuromethods, vol. 171,
https://doi.org/10.1007/978-1-0716-1589-8_5, © Springer Science+Business Media, LLC, part of Springer Nature 2022

(i.e., autophagy induction) and impaired lysosome fusion (i.e., autophagy inhibition).

In contrast, assays that measure autophagic flux provide a readout for overall autophagic activity or, in other words, flow through the entire autophagy process. Such assays work by following a labeled substrate of autophagy until it is degraded by the lysosome. For example, the metabolic pulse-chase (MPC) assay measures autophagic flux by labeling a pool of proteins with a radiolabeled amino acid and then measuring the decreasing levels of labeled proteins over time. Critically, the "pulse" phase of the experiment in which proteins are labeled is separated in time from the "chase" phase in which protein levels are measured. This experimental design allows the measurement of changes in clearance rate without potentially confounding effects from changes in protein synthesis.

Recent advances in the development of photoconvertible fluorescent proteins (PCFP), such as Dendra or EOS, and time-lapse microscopy have enabled our lab to develop a new approach to measuring flux we call optical pulse labeling (OPL), which is analogous to the MPC assay. OPL relies on the expression of an autophagy substrate, typically LC3, tagged with a PCFP, which undergoes irreversible green-to-red conversion upon exposure to light of a specific wavelength, typically UV. This photoconversion acts as the "pulse" that converts a pool of LC3-EOS from green to red. During the "chase" period, the decline in red signal intensity is measured by time-lapse microscopy at regular intervals over subsequent hours or days. The rate of red signal decline can be converted to a protein half-life, which is a measure of autophagic flux.

The microscopy-based nature of OPL confers several key advantages over the MPC assay. OPL affords single-cell resolution, and protein half-lives can be calculated on a per-cell basis. This resolution enables the parsing and comparison of subpopulations of cells within a heterogeneous sample. In contrast, MPC is a population-based assay and is limited to providing the average half-life of proteins in all cells in a sample.

OPL is typically performed as a longitudinal experiment, whereby measurements are made in the same cells over time. In contrast, MPC is done by running multiple samples in parallel but collecting the samples at different chase intervals. The longitudinal approach enabled by OPL has the advantage of achieving high sensitivity from relatively few samples. Since each cell effectively serves as its own control, biological and technical variability are better managed. The relatively small number of cells needed for OPL as compared to MPC also makes it amenable to high-throughput screening applications.

Another advantage of OPL is the ability to multiplex it with other measurements or reporters, enabling the collection of very rich datasets. Since photoconversion is always incomplete and protein synthesis of new reporter continues after photoconversion, the morphology of each cell remains visible in the green channel throughout the experiment. Cell morphology contains rich information related to cell health and function, which can be analyzed alongside autophagic flux. Co-expression of an OPL reporter with other fluorescent markers or reporters can provide even more structural or functional information. These additional measurements can then be correlated to autophagic flux to provide deeper mechanistic insight. For example, co-expression of an OPL reporter with a fluorescently tagged disease protein could be used to investigate the relationship between autophagic flux and disease protein aggregation.

An important caveat that applies to either OPL or MPC is the potentially confounding effects of clearance by other systems, such as the ubiquitin-proteasome system (UPS). Technically, OPL and MPC measure the overall clearance rate, but do not directly provide information about which system or systems are mediating clearance. Commonly, autophagy flux measurements are conducted in the presence of a proteasome inhibitor to isolate the role of autophagy [3], although evidence suggests that inhibition of the UPS may itself induce autophagic clearance of some substrates [4, 5]. OPL with an LC3-PCFP reporter may more directly interrogate autophagic clearance since LC3 is consumed during autophagy. Note that an LC3-PCFP reporter would not be useful for measuring the rate of chaperone-mediated autophagy or forms of macroautophagy, which are LC3-independent [6, 7], but OPL could, in principle, be used to measure flux of these processes using a different fusion protein.

Indeed, there may be no assay that specifically measures autophagic flux under all conditions. Given this limitation, we recommend following up any flux assay with several assays that probe the mechanics of clearance (reviewed elsewhere in this volume), such as a western blot analysis for LC3 lipidation. In this way, a flux assay is used to determine the presence and direction of an effect on autophagic clearance, and complementary assays are used to probe the mechanisms that underlie the effect.

This chapter provides a detailed protocol for conducting an OPL autophagic flux assay in primary neuronal cultures.

2   Materials

2.1   Primary Neuron Prep and Culture

Primary cortical neurons are dissected from E17 to 18 pups from timed pregnant female CD-1 mice or Sprague-Dawley rats (Charles River Laboratories, Wilmington, MA), as previously described [8]. Dissociated neurons are plated at 600,000 cells per cm^2 on poly-D-lysine-coated 96- or 384-well plates (PerkinElmer,

Waltham, MA). Primary neurons are grown in Neurobasal medium without phenol red (to improve contrast in the red channel), supplemented with 2% B27, 1% penicillin/streptomycin, and 1% GlutaMAX (Gibco, Waltham, MA). Media should be changed every 10–14 days.

2.2 Photoconvertible Fluorescent Proteins and Expression System

Many new PCFPs have been discovered in recent years, and new variants with improved properties continue to be developed [9–11] (*see* **Note 1**). Our lab has had success with either Dendra2 or mEOS3.2 [9, 11] (*see* **Note 2**). We typically prefer mEOS3.2 for its superior brightness and fast maturation rate [9].

Dendra2 and mEOS3.2 can function well as N- and C-terminus tags, but the optimal tag location should be determined empirically for any given protein of interest. For LC3, we typically fuse the PCFP to the C-terminus. Since primary neuronal cultures are relatively pure, we prefer to use a CAG promoter to drive high expression levels. For applications where neuronal specificity is important, we use a synapsin-1 promoter. Numerous Dendra2- and mEOS3.2-containing plasmids are available from Addgene. We typically express OPL reporters by transient transfection with Lipofectamine 2000 (Invitrogen, Carlsbad, CA), but other overexpression systems can also be used.

2.3 Positive Controls

When assessing the impact of various perturbations on autophagic flux, it is important to compare them to known modulators of autophagy. Also, these modulators are useful when optimizing and validating the specific experimental conditions to be used in an OPL experiment. A variety of small molecules and genes are known to modulate autophagic flux and can be used as positive controls.

To reduce autophagic flux, we use bafilomycin A1 (BafA), which blocks autophagosome-lysosome fusion and/or lysosome acidification [12, 13]. We treat with relatively low concentrations of BafA (25–100 nM) for up to 48 h, but the optimal concentration and incubation time should be empirically determined to balance autophagy inhibition with toxicity.

We have used several approaches to increase autophagic flux, the choice of which depends on specific experimental requirements. A commonly used method is to starve the cells by incubating them in serum-free media for several hours [14, 15]. But OPL measures clearance over 12–36 h and hence would require a relatively long period of starvation, which is toxic to primary neurons. Instead, we prefer to use small molecules or siRNAs to induce autophagy, as they are more compatible with cell survival over the duration of OPL experiments.

Our lab recently identified several small molecules that induce autophagy and can be used as positive controls in an OPL experiment: fluphenazine (FPZ), methotrimeprazine (MTM), and

10-(4'-(N-diethylamino)butyl)-2-chlorophenoxazine   (10-NCP) [16, 17]. All three small molecules can be used to treat primary neurons at 0.5 µM for several days with no overt toxicity. We have also had success with 0.5 µM rapamycin, which inhibits the activity of MTOR, a negative regulator of autophagy [18].

For experiments that test the effect of siRNAs on autophagy, it is often desirable to use an siRNA as a positive control. In this case, we have found that co-transfection of *Mtor* or *Rubicon* siRNA (10–50 nM) induces autophagic flux, although less strongly than the small molecules listed above (unpublished observations).

2.4 Automated Fluorescence Microscopy

To image neurons longitudinally requires a fluorescent microscope equipped in a way that automates as much of the imaging as possible. At a minimum, OPL requires a fluorescence microscope with an environmental chamber and a motorized stage with the capability of returning to the same imaging positions at each timepoint.

In our lab, we use fluorescence microscopes equipped to fully automate the imaging process, as previously described [8, 19]. With this setup, we can run multiple OPL experiments in sequence, which is useful for high-throughput applications. We use a Nikon Eclipse Ti-E inverted microscope with an ASI MS-2000 motorized stage, Nikon Perfect Focus System, high numerical aperture 20× objective, Andor Zyla 4.2 sCMOS, all enclosed within an environmental chamber set to 37 °C and 5% CO_2. Illumination is provided by a Sutter Lambda XL Xenon arc lamp with filters for DAPI, FITC, and TRITC. Experiments on 96- or 384-well polystyrene microwell plates are loaded into a Liconic STX44-ICBT robotic incubator and transferred to the microscope stage by a PAA KinedX articulated robotic arm.

Automation of these components is controlled by a mix of custom and proprietary software through Micro-Manager [20]. For each experiment, all imaging parameters, including exposure settings and the imaging interval, are preloaded into the microscope job handler. The position of a microwell plate on the microscope will shift slightly between timepoints, so to achieve good registration between timepoints, it is essential that the microscope have a system for imaging the same fields each time. For example, our automated microscopes run a custom routine that takes the top-left corner well as a fiducial marker, calculates an offset for each timepoint relative to the first timepoint, and uses the offset to realign the plate between timepoints.

2.5 UV Photoconversion Source

mEOS3.2 and Dendra2 are most efficiently photoconverted in the UV range (~400 nm) [9, 11]. For low-throughput applications, photoconversion can be performed on the microscope one field at a time using DAPI illumination. For high-throughput applications, we use a custom-designed UV light box to photoconvert all wells of a microwell plate simultaneously. This light box contains an array of closely spaced SMD LEDs that emit at 390 nm, enclosed within a black Plexiglas box, the top of which has a plate holder located 3.5 cm above the LEDs. This LED array emits UV light at approximately 30 mW/cm^2 at the plate height. Alternatively, we have had success using a Loctite LED Flood Controller with EQ CL30 LED Flood 405 lamp (emits at approximately 800 mW/cm^2 at 5 cm), which will photoconvert approximately three-quarters of a multi-well plate at once. The UV light emitted from these sources is very powerful, so it is important to wear UV eye protection and minimize skin exposure.

2.6 Image Processing and Analysis Software

Single-cell analysis of autophagic flux relies on the accurate segmentation and tracking of individual neurons over time. Automated cell segmentation and tracking is an area of active investigation with new methods and tools emerging regularly. Our lab uses a pipeline of custom Python-based scripts, which perform background subtraction, adaptive threshold-based segmentation, cross-timepoint alignment, proximity-based tracking, and single-cell intensity measurements. However, this pipeline can also be created using free image analysis packages such as ImageJ [21] or CellProfiler [22]. Once single-cell measurements have been obtained, custom scripts written in R (https://www.r-project.org/) are used to carry out data quality control filtering, model fitting, statistical analysis, and data visualization.

3 Methods

3.1 Expression of OPL Reporter in Primary Neurons

Primary neurons are isolated from embryonic mouse or rat pups and plated into 96- or 384-well microwell plates. Two days later, neurons are transfected with the OPL reporter. We typically transfect with 0.1–0.3 μg of reporter DNA per 100,000 cells (*see* **Note 3**). If using genetic manipulations as controls or test conditions (e.g., siRNA, overexpression plasmids), we co-transfect these with equimolar concentrations of the OPL reporter. As discussed below, green-to-red photoconversion is relatively inefficient at nontoxic levels of UV, so it is important to express PCFPs as highly as possible. In addition, the neurons should be in media without phenol red to maximize contrast in the red channel.

If using pharmacological treatments as controls or test conditions, apply these before photoconverting. The timing of such treatments will depend on how quickly the drugs are expected to act on clearance. For autophagy inhibitors such as BafA, we typically treat 4 h prior to photoconversion. For autophagy inducers such as FPZ, MTM, and 10-NCP, we typically incubate for 12–24 h prior to photoconversion.

3.2 UV Optimization   UV exposure time and intensity should be optimized empirically to balance red channel brightness with UV toxicity (Fig. 1, *see* **Note 4**). UV exposure typically falls between 45 and 90 s at 30 mW/cm². Photoconversion should result in a red channel signal that is clearly visible in the soma, with only faint labeling of neurites. The method described herein measures autophagic flux in the soma only, so it is not necessary to induce visible red signal in the neurites. In our

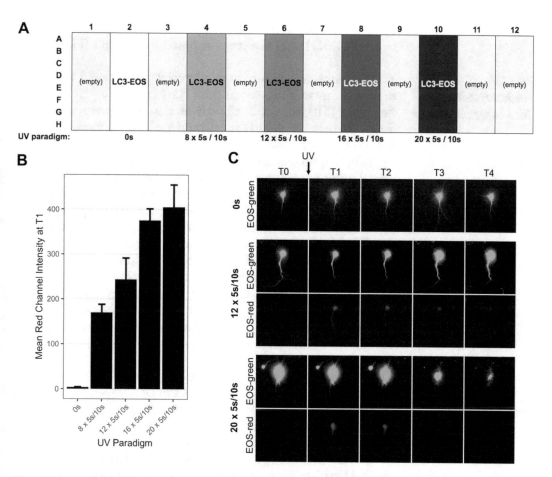

Fig. 1 Representative UV exposure optimization experiment. The optimal UV exposure should be determined empirically to find a balance between good photoconversion and minimal toxicity. Test conditions should be arranged in columns on a microwell plate, separated by empty columns to avoid UV light bleed-through from adjacent conditions (**a**). Each total UV duration should be broken up into pulses to help minimize UV-related toxicity, e.g., eight 5 s pulses with 10 s between pulses = 40 s of cumulative UV. A condition with no UV (0 s) should be included to establish a baseline for neuron survival. Each column should be exposed to UV light individually, while other columns are blocked with aluminum foil. Mean EOS-red intensity at the first timepoint following UV exposure (T1) was plotted and revealed a dose-response to increasing UV exposures (**b**). Images should be inspected to determine whether each UV condition results in toxicity. In this example, 20 × 5 s/10 s of UV resulted in excellent photoconversion, but caused many neurons to die within a few timepoints (**c**). 12 × 5 s/10 s of UV was found to give the optimal balance of good photoconversion and low toxicity

experience, a UV exposure that produces bright red signal in neurites is more likely to cause overt toxicity in the hours or days following UV exposure. In this case, the UV intensity or exposure time should be reduced.

3.3 Imaging Optimization

Imaging should begin once reporter expression has peaked, typically 2–3 days after transfection. Since there is no signal in the red channel prior to photoconversion, we recommend including a small number of extra wells that can be used to determine optimal acquisition settings before photoconverting the whole plate (*see* **Note 5**). During these tests, aluminum foil can be used to shield any wells that should not yet be exposed to UV light (Fig. 1).

3.4 UV Photoconversion and Time-Lapse Imaging

Prior to photoconversion, each microwell plate should be imaged in both the green and red channels (Fig. 2a). Although this pre-UV timepoint (T0) will not be directly used in the analysis, it is useful for assessing the health of the cells and the baseline levels of (auto)-fluorescence in the red channel.

After T0 images are acquired, all wells should be photoconverted using the optimized UV intensity and duration. Photoconversion is essentially instantaneous after UV exposure, so collection of post-UV (T1) images should begin promptly to capture the red signal at its peak. After this and subsequent timepoints, the plate should be returned to an incubator to maintain optimal growth conditions. The plate should return to the microscope to be imaged using the same imaging parameters at regularly spaced intervals for 12–36 h. For OPL with LC3 ($t_{1/2} = 15$ h [16]), we typically collect 4–5 post-UV timepoints spaced every 3–6 h (*see* **Note 6**). The exact timing of each timepoint should be recorded since it will be used in the analysis to model the red signal decline.

3.5 Image Processing and Measurement

The precise methods for processing images depend on the specific nature of the image acquisition. Generally, images should be background-corrected to reduce vignetting and shading artifacts, individual tiles of a well should be stitched together to form a montage, and montages should be aligned across time. Once images are processed, individual neuronal soma should be segmented based on the green channel, which represents the morphology of each neuron, and neuron identity tracked across time. We use a custom Python-based pipeline for this, but processing can usually be done using modules within the microscope software or freely available software such as ImageJ and CellProfiler. Regardless of the tool used, it is critical that neuron tracking be checked for accuracy, and that each neuron be checked for survival for the duration of the imaging time course. Neurons that are mis-tracked or die should be excluded from the analysis. Once

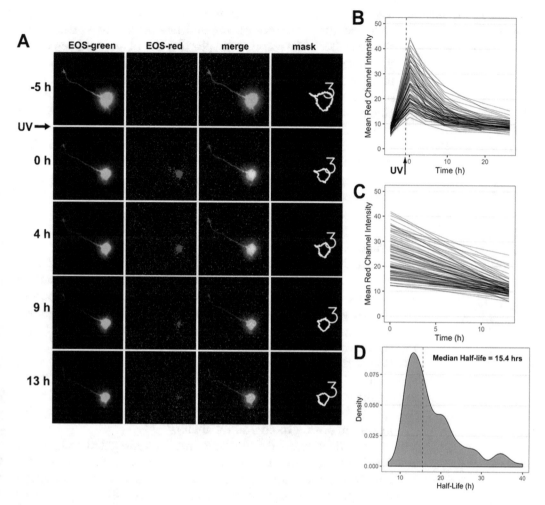

Fig. 2 Representative OPL experiment with LC3-EOS overexpressed in mouse primary neurons and imaged by automated longitudinal microscopy. Prior to UV exposure (−5 h), only autofluorescence is detectable in the red channel (**a**, **b**). Exposure to UV converts a subset of EOS-green to EOS-red, which represents a labeled pool of LC3. The clearance rate of this pool of LC3 is determined by imaging the same positions at regular time intervals. Individual neurons are segmented based on EOS-green+ soma and assigned an identification number (e.g., "3," in **a**), and mean EOS-red intensity is measured for each neuron within the segmentation mask at each timepoint (**b**). The EOS-red intensity curves are inspected to determine which timepoints should be included in the analysis, i.e., which timepoints after UV exposure are above the noise floor in the red channel. In this example, the EOS-red intensities from 0 to 13 h for each neuron are used to fit a linear regression (**c**), which is then used to calculate single-cell half-lives. The distribution of the half-lives of all neurons can be visualized using a population density plot (**d**), which is helpful for revealing heterogeneity in neuronal clearance rate. In this example, the median half-life of LC3 (dashed line) was 15.4 h

neurons have been accurately segmented and tracked based on the green channel (Fig. 2a), the red channel mean pixel intensities within each segmentation mask should be measured at each time-point (Fig. 2b). This is the raw data that will be used to calculate single-cell half-lives.

3.6 Single-Cell Half-Life Analysis

The half-life analysis should be performed only using timepoints collected after photoconversion and before the red signal reaches the noise floor. We recommend plotting the blinded single-cell raw data over time to determine when the red signal intensity declines to a plateau at noise floor level for most neurons (Fig. 2b). For each neuron, there should be at least three timepoints where the red channel signal is above the noise floor (*see* **Note 7**), otherwise the neuron should be excluded.

On a single-cell basis, red channel mean pixel intensities from the timepoints selected above are log-transformed and used to fit a linear regression (Fig. 2c). The half-life of the tagged protein is calculated for each neuron according to the following formula:

$$t_{1/2} = -\ln{(0.5)}/m$$

where m is the slope of the linear model.

At this point, the data should be cleaned to remove neurons with anomalously low or high half-lives. This can be done by plotting or sorting a blinded list of all half-lives, then inspecting the images that correspond to very low or very high half-lives. Typically, anomalous half-lives are the result of instances of poor segmentation, tracking, dead neurons, or imaging artifacts. Inspection of the data should reveal a range of half-lives, which should be included in the analysis. This process is also partly informed by the expected half-life of the tagged protein, if known, although one should be careful not to constrain the data so much that it eliminates interesting biological variability.

The final, cleaned dataset can then be aggregated and plotted. For high-throughput studies, we typically calculate the median half-life per condition. Hits can also be analyzed further by plotting the population density curve to visualize the distribution of half-lives in each condition (Fig. 2d).

4   Notes

1. Several characteristics should be considered when selecting a PCFP for OPL. An ideal PCFP should be bright (i.e., high quantum yield and extinction coefficient, and fast maturation time) and efficiently photoconvertible to minimize the UV exposure required to produce sufficient red signal. The photoconversion should be irreversible to ensure that a measured decrease in red signal is due to clearance only and is not confounded by photoswitching back to a green-emitting state. Also, an ideal PCFP should be monomeric as oligomerization might disrupt the function of the labeled protein of interest (e.g., LC3).

2. An important consideration is that Dendra2 is photoconverted by blue light, albeit less efficiently than by UV, which can lead to additional, unwanted photoconversion every time the green channel is imaged [23]. This effect can be mitigated by short exposure times and/or an attenuated blue light source. Indeed, this effect can be desirable when using a cell type that is particularly sensitive to UV, in which case blue light can be used instead of UV for photoconversion.

3. The amount of reporter DNA used should be empirically determined to maximize reporter brightness without inducing toxicity related to overexpression. Transfection efficiency should be relatively low so that labeling is sparse (<300 labeled cells per 100,000 cells). If labeling is too dense, it will be difficult to achieve accurate single-cell tracking. We recommend doing a pilot experiment with various combinations of plasmid concentration and UV exposure to determine the conditions that optimally balance red channel brightness with toxicity.

4. We recommend conducting a pilot experiment that tests multiple UV conditions on the same microwell plate. We do this by wrapping the plate in aluminum foil or a similarly opaque material, and then unwrapping individual columns of wells to be exposed to different durations of UV and/or UV sources. We have found that some UV light will leak under the edge of the foil, so it is best to skip one or two columns between conditions (Fig. 1a). It is also helpful to include a set of wells that is not exposed to UV, which can be used to disambiguate whether any toxicity is related to reporter overexpression or UV exposure. We have found that pulsing the UV light (e.g., 5 s pulses with 10 s between pulses) helps to reduce toxicity, possibly by mitigating UV-related temperature increases [24].

5. Exposure times should be just long enough to detect the signal clearly, but not so long as to cause excessive phototoxicity. The precise duration will vary depending on the illumination intensity and light path of the microscope being used. The exposure time for the green channel can be relatively short since the green state of mEOS3.2 and Dendra2 is bright, and the green signal is used only to track the morphology of neurons. Importantly, green-to-red photoconversion of mEOS3.2 or Dendra2 will be incomplete at nontoxic levels of UV exposure. Thus, PCFP signal in the red channel will be much dimmer than the green channel signal and will typically require an exposure time 2–4 times longer. Since half-life is calculated based on a decline in red signal intensity, it is important to set a long enough exposure time for the red signal to remain above the noise floor for at least several timepoints. Given that the red signal is relatively dim, it is helpful to use a microscope

equipped with a high-NA objective and high-sensitivity camera (e.g., back-illuminated sCMOS).

6. OPL tagging a protein with a shorter half-life than LC3 may require more frequent imaging. Conversely, OPL with longer-lived proteins may benefit from less frequent imaging that can be carried out over multiple days. Importantly, the red channel signal should be above the noise floor in all cells for at least three timepoints to make the analysis reliable.

7. If there are too few timepoints where the red channel signal is above the noise floor, then the experiment should be repeated with one or more of the following changes: (1) increase the amount of plasmid during transfection to increase the amount of PCFP available for photoconversion, (2) increase the UV exposure intensity and/or duration to increase the initial red channel intensity, (3) increase the frequency at which timepoints are collected, or (4) use a camera with greater sensitivity.

References

1. Klionsky DJ, Abdelmohsen K, Abe A et al (2016) Guidelines for the use and interpretation of assays for monitoring autophagy (3rd edition). Autophagy 12:1–222. https://doi.org/10.1080/15548627.2015.1100356

2. Finkbeiner S (2019) The autophagy lysosomal pathway and neurodegeneration. Cold Spring Harb Perspect Biol 12(3):a033993. https://doi.org/10.1101/cshperspect.a033993

3. Sha Z, Zhao J, Goldberg AL (2018) The ubiquitin proteasome system, methods and protocols. Methods Mol Bio (Clifton, NJ) 1844:261–276. https://doi.org/10.1007/978-1-4939-8706-1_17

4. Zhu K, Dunner K Jr, McConkey D (2009) Proteasome inhibitors activate autophagy as a cytoprotective response in human prostate cancer cells. Oncogene 29:onc2009343. https://doi.org/10.1038/onc.2009.343

5. Wang D, Xu Q, Yuan Q et al (2019) Proteasome inhibition boosts autophagic degradation of ubiquitinated-AGR2 and enhances the anti-tumor efficiency of bevacizumab. Oncogene 38 (18):3458–3474. https://doi.org/10.1038/s41388-019-0675-z

6. Engedal N, Autophagy SP (2016) Autophagy of cytoplasmic bulk cargo does not require LC3. Autophagy 12(2):439–441. https://doi.org/10.1080/15548627.2015.1076606

7. Szalai P, Hagen L, Sætre F et al (2015) Autophagic bulk sequestration of cytosolic cargo is independent of LC3, but requires GABARAPs. Exp Cell Res 333:21–38. https://doi.org/10.1016/j.yexcr.2015.02.003

8. Tsvetkov AS, Arrasate M, Barmada S et al (2013) Proteostasis of polyglutamine varies among neurons and predicts neurodegeneration. Nat Chem Biol 9(9):586–592. https://doi.org/10.1038/nchembio.1308

9. Zhang M, Chang H, Zhang Y et al (2012) Rational design of true monomeric and bright photoactivatable fluorescent proteins. Nat Methods 9:727. https://doi.org/10.1038/nmeth.2021

10. Moeyaert B, Bich N, Zitter E et al (2014) Green-to-red photoconvertible Dronpa mutant for multimodal super-resolution fluorescence microscopy. ACS Nano 8:1664–1673. https://doi.org/10.1021/nn4060144

11. Chudakov DM, Lukyanov S, Lukyanov KA (2007) Tracking intracellular protein movements using photoswitchable fluorescent proteins PS-CFP2 and Dendra2. Nat Protoc 2:2024–2032. https://doi.org/10.1038/nprot.2007.291

12. Klionsky DJ, Elazar Z, Seglen PO, Rubinsztein DC (2008) Does bafilomycin A1 block the fusion of autophagosomes with lysosomes? Autophagy 4:849–850. https://doi.org/10.4161/auto.6845

13. Yamamoto A, Tagawa Y, Yoshimori T et al (1998) Bafilomycin A1 prevents maturation of autophagic vacuoles by inhibiting fusion between autophagosomes and lysosomes in rat hepatoma cell line, H-4-II-E, cells. Cell Struct Funct 23:33–42. https://doi.org/10.1247/csf.23.33

14. Mejlvang J, Olsvik H, Svenning S et al (2018) Starvation induces rapid degradation of selective autophagy receptors by endosomal microautophagy. J Cell Biol 217 (10):3640-3655 https://doi.org/10.1083/jcb.201711002

15. Roscic A, Baldo B, Crochemore C et al (2011) Induction of autophagy with catalytic mTOR inhibitors reduces huntingtin aggregates in a neuronal cell model. J Neurochem 119:398–407. https://doi.org/10.1111/j.1471-4159.2011.07435.x

16. Barmada SJ, Serio A, Arjun A et al (2014) Autophagy induction enhances TDP43 turnover and survival in neuronal ALS models. Nat Chem Biol 10:677–685. https://doi.org/10.1038/nchembio.1563

17. Tsvetkov AS, Miller J, Arrasate M et al (2010) A small-molecule scaffold induces autophagy in primary neurons and protects against toxicity in a Huntington disease model. Proc Natl Acad Sci U S A 107:16982–16987. https://doi.org/10.1073/pnas.1004498107

18. Jung C, Ro S-H, Cao J et al (2010) mTOR regulation of autophagy. FEBS Lett 584:1287–1295. https://doi.org/10.1016/j.febslet.2010.01.017

19. Arrasate M, Mitra S, Schweitzer ES et al (2004) Inclusion body formation reduces levels of mutant huntingtin and the risk of neuronal death. Nature 431:805. https://doi.org/10.1038/nature02998

20. Edelstein AD, Tsuchida MA, Amodaj N et al (2014) Advanced methods of microscope control using μManager software. J Biol Methods 1:10. https://doi.org/10.14440/jbm.2014.36

21. Schindelin J, Rueden CT, Hiner MC, Eliceiri KW (2015) The ImageJ ecosystem: an open platform for biomedical image analysis. Mol Reprod Dev 82:518–529. https://doi.org/10.1002/mrd.22489

22. Carpenter AE, Jones TR, Lamprecht MR et al (2006) CellProfiler: image analysis software for identifying and quantifying cell phenotypes. Genome Biol 7:R100. https://doi.org/10.1186/gb-2006-7-10-r100

23. Chudakov DM, Lukyanov S, Lukyanov KA (2007) Using photoactivatable fluorescent protein Dendra2 to track protein movement. BioTechniques 42:553, 555, 557 passim. https://doi.org/10.2144/000112470

24. Peixoto H, Moreno R, Moulin T, Leão RN (2018) Modeling the effect of temperature on membrane response of light stimulation in optogenetically-targeted neurons. PeerJ Preprints 6:e27248v1. https://doi.org/10.7287/peerj.preprints.27248v1

Chapter 6

Measuring Autophagosome Flux

André du Toit, Jan-Hendrik S. Hofmeyr, and Ben Loos

Abstract

Autophagy is an evolutionarily conserved catabolic process that plays an import role in cellular proteostasis. The continual degradation, and recycling, of portions of the cytoplasm through autophagy eliminates unused and toxic proteins and organelles, promoting a functional proteome and cellular function. Autophagy also serves as an important adaptive mechanism that protects against metabolic perturbation. Loss of autophagy activity has major detrimental effects and has been shown to lead to neuronal proteotoxicity, protein aggregation, and cell death onset associated with neurodegeneration. Studies aimed at modulating autophagy activity have shown promising results in clearing toxic protein cargo and preserving neuronal viability. Neurons are characterized by a particularly efficient autophagy system. However, to finely control autophagy activity requires the precise and accurate measurement of the autophagy flux, i.e., the rate of flow along the entire autophagy pathway. Fluorescence microscopy has substantially contributed to the assessment of autophagy, due to its ability to identify the autophagy pathway intermediates, and to describe them kinetically, in the entire cell volume. However, the quantitative discernment between autophagy pathway intermediates, particularly the autophagosome pool size and the autophagosome flux, has remained challenging. Here, we describe a single-cell analysis approach that allows the characterization of the autophagy system in terms of the pathway intermediate steady-state pool size, the autophagosome flux, and the transition time.

Keywords Autophagy, Autophagosome flux, Fluorescence microscopy, Vesicular pathway, Steady-state variables

1 Introduction

Macroautophagy, hereafter referred to as autophagy, is a highly conserved catabolic process that plays a vital role in the homeostatic maintenance of a functional proteome, which is crucial for cellular function. The autophagy pathway is tightly anchored in an energetic feedback loop that, upon nutrient depletion, begins with the formation of the characteristic isolation membrane/phagophore, which encapsulates bulk cytoplasmic cargo in a double lipid membrane [1]. Once the vesicle is elongated and fully matured, which is then called an autophagosome, it is transported dynein-dependently along the tubulin network toward the perinuclear

Ben Loos and Esther Wong (eds.), *Imaging and Quantifying Neuronal Autophagy*, Neuromethods, vol. 171,
https://doi.org/10.1007/978-1-0716-1589-8_6, © Springer Science+Business Media, LLC, part of Springer Nature 2022

region where it fuses with and delivers its cargo to lysosomes for degradation [2]. The continual removal of cytoplasmic proteins and damaged organelles through autophagy, and the concomitant replenishment thereof, maintains a functional proteome, contributes to organelle quality control while minimizing the buildup of unused, nonfunctional, and toxic proteins and organelles. Furthermore, autophagy serves physiologically as a critical adaptive mechanism that protects against metabolic perturbation; increasing cytoplasmic turnover to supply amino acids for energy production during periods of starvation. In this manner, autophagy activity, i.e., the rate of autophagic degradation, or autophagic flux, fine tunes cellular metabolite supply to its metabolic demand. Deviation in autophagy activity may have detrimental effects and has been linked to the onset and progression of several pathologies, most notably neurodegeneration. Studies have shown promising results for treating neurodegenerative diseases by modulating autophagy to clear toxic protein aggregates [3]. In order to understand the physiological and pathological roles of autophagy and develop effective autophagy modulating therapies requires the accurate measurement of autophagy activity. A number of techniques have been developed to assess autophagy activity by measuring the abundance or activity of key autophagy machinery proteins [4].

One of the simplest approaches to assess autophagy is by measuring the abundance of proteins involved in autophagy. Microtubule-associated proteins 1A/1B light chain 3, LC3, has been widely used in this context. Cytoplasmic LC3 protein undergoes a series of biochemical transformations that result in its conjugation to phosphatidylethanolamine, forming LC3-II, which is incorporated into the autophagosomal membrane. The level of LC3-II can be used as an indicator of autophagosomes abundance. The rate of autophagosome turnover can be assessed by measuring the difference in LC3-II levels in the presence and absence of autophagosome/lysosome fusion inhibitors, which reflects the amount of LC3-II delivered to lysosomes for degradation [5]. While biochemical assays of autophagy proteins can indicate whether or not autophagy activity increases, it is less suitable for the precise quantification of the rate of autophagosome turnover. Western blotting in particular, often presents with major variability and background, making it difficult to assess small changes in LC3 protein levels accurately. Moreover, and may be more important, western blotting does not measure a rate. Fluorescence microscopy, however, enables the visualization of autophagosomes by fluorescent labeling of LC3, typically with green fluorescent protein (GFP) that allows for the numerical quantification autophagosomes in live cells over time. Optical sectioning through the whole cell allows to report the entire autophagosome pool. However, dynamic organelle and membrane rearrangements are characteristic of the autophagy pathway, where synthesis, fusion, degradation, and recycling events impact the abundance of autophagy pathway intermediates. Tandem fluorescent constructs, such as RFP-GFP-LC3 (tandem fluorescent-tagged LC3, tfLC3) have enabled the identification and quantification of the intracellular autophagosome and autolysosome pool [6]. Since autophagosome pool size does not infer autophagy flux, both pool size and flux parameters are required. The fluorescent-based flux probe developed by Kaizuka et al. [7] allows rapid visualization of autophagy activity based on the abundance of GFP-LC3 and RFP-LC3ΔG following the proteolytic cleavage of GFP-LC3-RFP-LC3ΔG by Atg4. While GFP-LC3 undergoes lipidation and incorporation into the autophagosomal membrane where it is subsequently degraded by lysosomal proteases, RFP-LC3ΔG, which is unable to undergo lipidation, resides stably in the cytoplasm, thereby serving as internal control. Thus, the ratio of GFP/RFP signal can be used as a measure of autophagy degradative activity. This approach allows for the rapid screening of autophagy markers and has enabled most powerful and rapid visualization of autophagic flux and its differential distribution in tissues, such as spinal cord [7]. However, the ratiometric signal is less able to reflect intracellular pool sizes of autophagy pathway intermediates. Moreover, utilizing Atg4 as an indicator of autophagy activity is not without its limitations, as it may not directly reflect on the autophagy pathway and its activity, since Atg4 activity and abundance may be regulated at transcriptional and posttranslational level [8] as well as the intracellular microenvironment [9].

It is known that cells are characterized by an inherent autophagic activity, and it is important to discern that high or low autophagosome abundance does not infer high or low autophagic flux [10]. Particularly, neurons may present with low autophagosome pool sizes but high flux, while dysfunctional neurons may present with high pool size but low flux, suggesting that the relationship between flux and pool size requires attention.

Our laboratory has developed a fluorescence-based single-cell analysis approach that allows for the characterization of the autophagy system in terms of the pathway intermediates' steady-state (i.e., when synthesis rate and degradation rate are equal) pool size, the autophagosome flux (J), i.e., the rate of flow along the autophagy pathway, and the transition time (τ), indicating the time required by the cell to clear its entire autophagosome pool. This approach relies on the ability to accurately distinguish and monitor autophagy pathway intermediates/vesicles over time and to calculate autophagy steady-state properties based on a transient time-dependent profile. In this chapter, the experimental procedure is described for the quantitative assessment of the total number of autophagosomes (n_A), autolysosomes (n_{AL}), and lysosomes (n_L) over time in the presence and absence of an autophagosome/lysosome fusion inhibitor for measuring J, as the rate of autophagosome turnover at steady-state under basal and rapamycin-induced conditions. This procedure allows to accurately determine basal autophagic flux in various neuronal cell types and to discern the response to autophagy-enhancing drugs in a precise, sensitive, and quantifiable manner.

2   Materials

2.1   Cell Culture

Mouse hypothalamic GT 1–7 cells and mouse embryonic fibroblast (MEF) cells (a kind gift from Noboru Mizushima, Tokyo University) that stably express green fluorescent protein tagged to light chain 3 (GFP-LC3) were maintained in Dulbecco's modified Eagle's medium (DMEM) (Life Technologies, #41-965-039), supplemented with 10% fetal bovine serum (Biochrom, #S-0615) and penicillin-streptomycin (Life Technologies, #15-140-122), at 37 °C in a 5% CO_2 atmosphere.

2.2   Chemicals

Rapamycin (Sigma-Aldrich, #R-0395) was used as an autophagy inducer, and bafilomycin A_1 (LKT Laboratories Inc., #B-0025) as an autophagosomes/lysosomes fusion inhibitor, to completely block fusion between autophagosomes and lysosomes [11].

2.3   Image Acquisition

Cells were seeded onto CYTOO micro-patterned slides with large fibronectin disc shapes (CYTOO, #10-003-10) and allowed to adhere. Nonadherent cells were washed off after 15–20 min, and remaining adhering cells were maintained with culture medium supplemented with LysoTracker red (Thermo Fisher Scientific, #L-7528) for 6 h before imaging. Fluorescence-based live-cell imaging was performed on an Olympus IX81 wide-field microscope equipped with a stage incubator that was maintained at 37 °C in a 5% CO_2 atmosphere. Images were acquired using a 60× oil immersion objective and automated z-stack/stage control to acquire micrographs with a 0.5 μm step-width between image frames. Image stacks were processed and deconvolved using $Cell^R$ and analyzed with ImageJ/Fiji [12] to quantify the complete pool size of autophagy pathway intermediates (n_A, n_{AL}, n_L) over time, using a minimum of 10 cells.

3 Methods

3.1 Measuring Autophagy Variables

The experimental procedure to characterize the autophagy steady-state variables involves measuring the total autophagy intermediate abundance over time and calculating J as the initial rate of accumulation of n_A after the complete inhibition of the fusion between autophagosomes and lysosomes at steady state.

3.2 Flux Is the Rate of Flow Through the Autophagy Pathway at Steady State

To measure autophagosome flux requires that the autophagy system is in a steady state. This can be verified by imaging cells 1 h apart and then quantifying the total pool size n_A, n_{AL}, and n_L per cell. The autophagy system is in steady state when n_A remains constant over time for at least three time points, which indicates that the rate of autophagosome synthesis equals the rate of autophagosome degradation. The average number of n_A, n_{AL}, and n_L over the steady-state period is the autophagy steady-state pathway intermediates concentration (see **Note 1**). If there is significant variation in the n_A over time, it indicates that the autophagy system is in a transition state and, therefore, should be continuously monitored until a steady state has established before proceeding to the next step. For example, a cellular system is in a transition state when autophagy is induced, where synthesis rate increases. When the autophagy system is in a steady state, the fusion of autophagosomes and lysosomes is experimentally inhibited, and J is calculated from the initial slope of the increase in n_A at the point of inhibition of fusion. Note, by using smaller intervals (e.g., 30 min) for assessing, J improves its accuracy as it limits the effect of feedback mechanism.

3.3 The Complete Inhibition of Fusion Between Autophagosomes and Lysosomes

In order to measure autophagosome flux J, it is important that a complete inhibition between autophagosomes and lysosomes is achieved (see **Note 2**). Incomplete fusion inhibition will result in an inaccurate measurement of flux, since a residual autophagosome degradation rate, may remain. Hence, the concentration of fusion inhibitor required for the complete inhibition of fusion should be determined for each cell and inhibitor type used [11]. The concentration of fusion inhibition that is most desirable is the concentration when there is no further increase in the initial rate of n_A accumulation upon increasing fusion inhibitor concentration.

Figure 1 shows a cartoon of the transient time-dependent behavior of autophagosomes, autolysosomes, and lysosomes under basal and 25 nM rapamycin-induced conditions, indicating the steady-state variables. These data can be used to derive

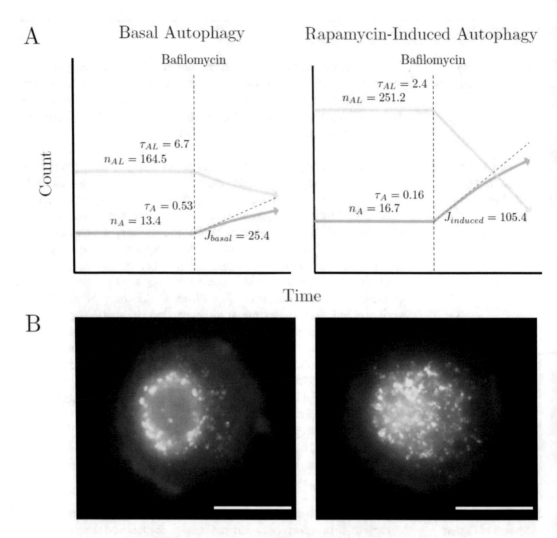

Fig. 1 Basal and rapamycin-induced autophagy variables. (**a**) A cartoon representation of basal and induced autophagy and the main derived variables (n_A autophagosome steady-state pool size; n_{AL} autolysosome steady-state pool size; τ transition time in h; J autophagosome/h/cell). (**b**) Representative micrographs of basal and rapamycin-induced steady-state autophagy. Scale bar: 25 μm. (Figure adapted from du Toit et al. [11])

valuable additional variables such as the transition time, which is the time required to clear the respective pool (Table 1). Additional control over intercellular variability can be achieved by using micropatterning, thereby enhancing data accuracy (*see* **Notes 3** and **4**). Care must be taken to carefully choose the acquisition time, so as to align temporal resolution with the vesicle dynamics at hand (*see* **Note 5**).

Table 1
Functional variables of autophagy for basal and rapamycin- (25 nM) induced autophagy in MEF cells (*A* autophagosomes, *AL* autolysosomes, *L* lysosomes)

Variable	Unit	Basal	Induced
Autophagosome flux, J	A/h/cell	25.4	105.4
Number of autophagosomes, n_A	A/cell	13	17
Number of autolysosomes, n_{AL}	AL/cell	165	251
Number of lysosomes, n_L	L/cell	1	1
Autophagosomal transition time, τ_A	h	0.53	0.16
Autophagolysosomal transition time, τ_{AL}	h	6.7	2.4

Derived variables are shown in italics. Table adapted form du Toit et al. [11]

4   Notes

1. *Distinguishing between the autophagy pathway intermediates* (n_A, n_{AL}, n_L).

   In order to generate accurate and reliable data of the autophagy pathway intermediates, their precise identification and subsequent quantification is required. A number of probes have been developed to fluorescently label autophagy vesicles. One of the simplest approaches is to fluorescently tag autophagy pathway proteins, such as GFP-LC3 and red fluorescent protein tagged to lysosome-associated membrane glycoproteins (RFP-LAMP). Another approach, which takes advantage of the change in pH upon lysosomal fusion, makes use of fluorescent tandem constructs that exhibit pH-dependent changes in the emission spectrum; mRFP-GFP-LC3 tandem fluorescent protein described by Kimura et al. [6] exploits the difference in the properties of the two fluorescent proteins (i.e., GFP fluorescence is quenched in the acidic lumen of lysosomes, while the pH-stable RFP continues to emit fluorescence), and the acid-stable fluorescent Keima probe which exhibits pH-dependent shifts in excitation and emission spectrum [13]. However, heterogeneous expression levels can result in a varying signal/noise ratio, which increases variability. Organelle-specific fluorescence dyes with inherent binding properties may here have the advantage of homogeneous staining with optimal signal/noise ratio in live cells without the need for cell transfection. There are several fluorescent dyes that have been developed to label the vesicles involved in the autophagy pathway, such as CYTO-ID R autophagy cationic amphiphilic tracer dye that is selectively incorporated into autophagosomes, and acidotropic dyes such as LysoTracker that selectively accumulate and emit fluorescence in acidic organelles such as lysosomes. The homogeneous labeling of autophagy vesicles exhibited by fluorescent dyes improves batch analysis when using automated counting algorithms.

2. *Inhibiting the fusion of autophagosomes and lysosomes.*

This is achieved using vacuolar-type H^+-ATPase inhibitors such as bafilomycin A_1 or chloroquine, which disrupt the maintenance of a low-pH environment in lysosomes and prevent the fusion of autophagosomes and lysosomes. Caution should be taken when using acidotropic dye that accumulates and fluoresces in acidic organelles, such as LysoTracker, as they will be affected by these fusion inhibitors. However, it is possible to generate reliable measurements of autophagosome flux using LysoTracker since its fluorescent signal is maintained for 2 h after bafilomycin A_1 treatment [11]. Specifically, we observed that concentrations of LysoTracker at 75 nM were well tolerated and remain sufficiently long in the vesicles to determine autophagosome flux accurately in MEF cells. Unlike western blotting, where typically time frames of 4–8 h of bafilomycin treatment are implemented, here, using live-cell imaging, only 30 min to 2 h are required. In fact, the initial slope of the progress curve is most accurately reflecting autophagosome flux, especially considering that changes in metabolites after 4–8 h of bafilomycin treatment can influence autophagy via feedback mechanisms. The concentration of LysoTracker and the time frame in which reliable data can be generated should be determined for the respective cell lines.

3. *Enhancing data accuracy and physiological model through micropatterning.*

Micropatterning is a tool that allows for the control cell size, its geometric constraints, and its position. There are a number of advantages associated with using micropatterning. First, it is a powerful tool to reduce cellular variability that stems from variation in cell size, shape, migration, and proliferation. Second, it allows to precisely maneuver acquisition parameters and stage control settings, since a precise localization of cells can be encoded in the stage coordinate settings, including acquisition parameters in the z-dimension. The standardized geometry makes vesicle acquisition highly accessible, allowing a more precise total pool size quantification. Third, it can be used to create more complex geometries or sample environments, such as "organ on a chip" models that would allow for the generation physiologically relevant data, for instance, neuronal networks [14] (Fig. 2). Here, standardization of geometric constraints allows the assessment of fluxes in specific subcellular regions. Micropatterned coverslips can be self-fabricated as described by Carpi et al. [15] and modified to suite specific

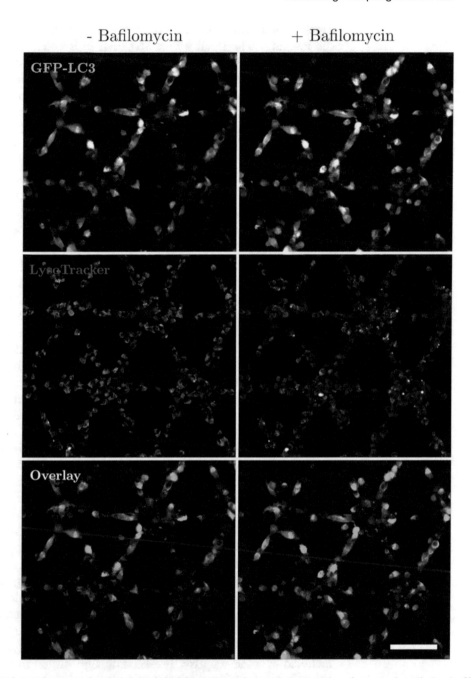

Fig. 2 Assessing autophagy activity in micro-patterned neuronal networks of mouse hypothalamic GT1–7 cells. The autophagy pathway intermediates and their pool size can be measured for individual cells, or, as shown here, the relative rate of autophagosome turnover can be assessed based on the change in GFP abundance in the absence and presence of 400 nM bafilomycin A1 treatment. This allows to interrogate autophagy activity according to the cellular arrangement, i.e., neuronal processes and interconnection vs. neuronal soma and nodes. Scale bar: 50 μm

experimental needs, and mounted to chamber slides for live-cell imaging. The longitudinal approach in live-cell imaging, combined with single-cell resolution to identify pathway intermediates with high accuracy, enables that each cell serves as its own control, which decreases variability further. Importantly, live-cell acquisition enables here to robustly determine, whether a steady state is present or whether the cell is in a transition state. In the latter, especially upon treatment with an autophagy inducer, this approach allows to assess whether the cell settles at a new, now heightened, steady state. These parameters become particularly important when discerning drug-induced or pathological flux deviation, or when a desirable change in flux requires particular dosing and scheduling [16].

4. *Image analysis.*

There are a number of techniques/algorithms available to count fluorescence puncta in an automated manner. While setting up an image analysis pipeline seems trivial, it may often be overlooked which can have unintended major negative impact on analysis accuracy, e.g., an image analysis pipeline performs well on selective micrographs that were used to set it up, but performs poorly when applied to batch analysis. In our experience, when developing a robust image analysis pipeline, three main factors should be considered: the detection algorithm, the mode of image transformation, and protein/organelle labeling strategies. A number of algorithms on various platforms (e.g., Scikit-image (Python) [17], OpenCV (Python, Java, C++) [18], ImageJ/Fiji [12]) exist with the relevant information describing the algorithmic foundation. The detection algorithm used would depend on the image topography, e.g., utilizing global threshold to detect puncta in high contrasting images. Where most pipelines fall short is during batch analysis to accuracy resolve puncta in micrographs or series/stacks with a varying signal-to-noise ratio. Preprocessing can greatly improve overall data analysis by applying image transformation algorithms. Deconvolution is a popular and powerful image transformation tool to enhance the signal-to-noise ratio. There are additional image transformation avenues that can be applied such as image smoothing/blur (typically included in some adaptive thresholding algorithms) to reduce the effects of artifacts on the accuracy of the algorithm. A simple way to improve images analysis is to select labels/dyes that produce high contrasting images. We have found that inherent binding probes are more often more desirable over plasmids that often exhibit heterogeneous expression and variable background noise. In such cases, the image analysis pipeline requires to be tested, to be best aligned for a given signal output.

5. *Experimental constraints/pitfalls.*

Several experimental/equipment factors can influence the accuracy when measuring autophagosome flux; temporal resolution, image acquisition, and resolution. Establishing whether or not autophagy is at steady-state is an important step in measuring autophagosome flux. Therefore, the temporal resolution, the time points used to measure the autophagy intermediates, must allow to sufficiently resolve the transient behavior of autophagy. This may not always be a trivial process. When transition time exceeds temporal resolution to the extent where a change in autophagy intermediates occurs on a time scale of days and data acquisition is on an hourly basis, it may seem that the autophagy system is in a steady-state while it is still in transition. Therefore, a fine balance between temporal resolution and transient properties of autophagy is necessary to accurately measure autophagy variables, which may require several attempts before establishing the optimal balance. Another aspect that requires consideration when measuring the complete autophagy intermediates pool is the acquisition time. Generating high-quality images suitable for analysis often requires a given acquisition time, which, when imaging rapidly moving vesicles, can result in artefacts being generated when moving the objective along the axial plane and switching between laser/filters. Therefore, image acquisition parameters must be chosen that ensure optimal signal while restricting artifacts. It must also be considered that this method is experimentally disruptive, as it blocks autophagosome/lysosome fusion, and allows a limited time window for acquisition. This protocol may, therefore, serve as a complementary tool, particularly suitable to assess and finely dissect to what extent autophagy-inducing drugs impact autophagosome flux.

Acknowledgments

This work was supported by the South African Medical Research Council (SAMRC), the South African National Research Foundation (NRF), and the Cancer Association South Africa (CANSA).

References

1. Loos B, Engelbrecht AM, Lockshin RA, Klionsky DJ, Zakeri Z (2013) The variability of autophagy and cell death susceptibility: unanswered questions. Autophagy 9 (9):1270–1285

2. Jahreiss L, Menzies FM, Rubinsztein DC (2008) The itinerary of from peripheral

formation to kiss-and-run fusion with lysosomes. Traffic 9:574–587

3. Ravikumar B, Vacher C, Berger Z, Davies JE, Luo S, Oroz LG, Scaravilli F, Easton DF, Duden R, O'Kane CJ, Rubinsztein DC (2004) Inhibition of mTOR induces autophagy and reduces toxicity of polyglutamine

expansions in fly and mouse models of Huntington disease. Nat Genet 36:585–595

4. Mizushima N, Yoshimori T, Levine B (2010) Methods in mammalian autophagy research. Cell 140:313–326

5. Mizushima N, Yoshimori T (2007) How to interpret LC3 immunoblotting. Autophagy 3:542–545

6. Kimura S, Noda T, Yoshimori T (2007) Dissection of the autophagosome process by a novel reporter protein, tandem fluorescent tagged lc3. Autophagy 3:452–460

7. Kaizuka T, Morishita H, Hama Y, Tsukamoto S, Matsui T, Toyota Y, Kodama A, Ishihara T, Mizushima T, Mizushima N (2016) An autophagic flux probe that releases an internal control. Mol Cell 17:835–849

8. Yoshimura K, Shibata M, Koike M, Gotoh K, Fukaya M, Watanabe M, Uchiyama Y (2006) Effects of rna interference of atg4b on the limited proteolysis of lc3 in pc12 cells and expression of atg4b in various rat tissues. Autophagy 2:200–208

9. Scherz-Shouval R, Shvets E, Fass E, Shorer H, Gil L, Elazar Z (2007) Reactive oxygen species are essential for autophagy and specifically regulate the activity of Atg4. EMBO J 26 (7):1749–1760

10. Loos B, du Toit A, Hofmeyr JHS (2014) Defining and measuring autophagosome flux—concept and reality. Autophagy 10 (11):2087–2096

11. du Toit A, Hofmeyr J-HS, Gniadek TJ, Loos B (2018) Measuring autophagosome flux. Autophagy 14:1060–1071

12. Schindelin J, Arganda-Carreras I, Frise E, Kaynig V, Longair M, Pietzsch T, Preibisch S, Rueden C, Saalfeld S, Schmid B et al (2012) Fiji: an open-source platform for biological-image analysis. Nat Methods 9:676–682

13. Katayama H, Kogure T, Mizushima N, Yoshimori T, Miyawaki A (2011) A sensitive and quantitative technique for detecting autophagic events based on lysosomal. Chem Biol 18:1042–1052

14. Hardelauf H, Waide S, Sisnaiske J, Jacob P, Hausherr V, Schöbel N, Janasek D, van Thriel C, West J (2014) Micropatterning neuronal networks. Analyst 139:3256–3264

15. Carpi N, Piel M, Azioune A, Fink J (2011) Micropatterning on glass with deep UV. Protoc Exch 10

16. Loos B, Klionsky DJ, du Toit A, Hofmeyr JHS (2020) On the relevance of precision autophagy flux control in vivo—points of departure for clinical translation. Autophagy 16 (4):750–762

17. Van der Walt S, Schönberger J, Nunez-Iglesias-J, Boulogne F, Warner J, Yager N, Gouillart E, Yu T (2014) scikit-image: image processing in python. PeerJ 2:e453

18. Bradski G, Kaehler A (2000) Opencv. Dr Dobb's J Softw Tools 120:122–125

Chapter 7

Measurement of Neuronal Tau Clearance In Vivo

Ana Lopez, Angeleen Fleming, and David C. Rubinsztein

Abstract

This chapter describes a new method to monitor and quantify tau protein clearance in vivo in living neurons of the zebrafish. The optical transparency of zebrafish allows visualization of fluorescently tagged proteins within cells and the evaluation of changes in protein levels. Here, we describe a fluorescent-based technique using the photoconvertible protein Dendra fused to tau to analyze the clearance rate of Dendra-tau protein in neurons in the spinal cord of living fish using confocal imaging. The intensity of photoconverted Dendra-tau, measured at 12-h intervals over a period of 2 days, reflects the clearance kinetics of the protein and can be used to investigate the effect of genetic modifiers or pharmacological modulators on protein turnover. The method described here focuses on the specific clearance of tau protein in neurons in the spinal cord, but it can easily be adapted to other Dendra-tagged proteins and different cell populations.

Keywords Zebrafish, Tau, Live imaging, Confocal microscopy, Protein clearance kinetics, Autophagy, Neuron, Photoconversion

1 Introduction

In recent years, there have been rapid advances in the technologies available for labeling autophagic vesicles, particularly the use of dual-color reporters and pH-sensitive probes (reviewed in [1]). These have been employed to measure the total number of autophagosomes and/or autolysosomes to infer how different conditions affect autophagic flux. An alternative approach is not to measure autophagosomes and autolysosomes but to measure the clearance of autophagic substrates. Traditionally, this has involved the use of biochemical pulse-chase approaches in vitro or inducible expression of constructs where expression of an autophagic substrate can be turned on or off by the addition of a compound to the medium (e.g., doxycycline to control expression from the Tet-on or Tet-off promoter, [2]). In the latter example, an autophagic substrate can be expressed for a given time period, and its clearance can be measured by collecting samples at defined time points after switch off, then analyzing the amount of substrate present relative to the amount at time zero, typically by western blotting. Such analysis is

Ben Loos and Esther Wong (eds.), *Imaging and Quantifying Neuronal Autophagy*, Neuromethods, vol. 171,
https://doi.org/10.1007/978-1-0716-1589-8_7, © Springer Science+Business Media, LLC, part of Springer Nature 2022

not only applicable to cell culture studies but has also been used to measure substrate accumulation and clearance in mouse transgenic models (e.g., tau clearance, [3]).

With the advent of photoconvertible fluorophores, such as mEOS and Dendra, it is now possible to perform such clearance analysis (termed optical pulse-chase) in live cells or in vivo. Dendra2 is a photoconvertible fluorophore with fast maturation and bright fluorescence, which can be irreversibly converted from green to red fluorescence by exposure to 405 nm light [4]. When linked to an autophagy substrate, photoconversion labels the existing pool of the substrate, the clearance of which can be measured by image analysis of the red fluorescence intensity. Such analysis is not confounded by the synthesis of new protein, since the newly formed protein will be labeled with the green (unconverted) fluorophore. Neurodegeneration-associated aggregate-prone proteins are ideal substrates for such analysis. Indeed, autophagy has been shown to be critical for the degradation of huntingtin, mutant α-synuclein, and tau [5–8], and understanding their clearance kinetics in living cells provides a powerful tool for assessing potential therapeutic compounds and genetic modifiers. This approach was first used by Tsvetkov et al. to study the clearance of Dendra-tagged huntingtin in cultured striatal neurons [9]. Our lab has since applied this approach in vivo, using zebrafish models expressing human tau tagged with Dendra2. We have used transient expression to assess genetic modifiers of tau clearance [10] and stable transgenic lines to assess clearance of wild-type and mutant forms of tau and their responses to autophagy stimuli [11]. Such studies have provided the first observations of substrate clearance in intact neurons in vivo and open the opportunity for spatial analysis of clearance kinetics (e.g., neuronal cell body, dendrites, synapse). The selection of the autophagy clearance substrate is an important consideration, since many substrates are likely cleared by both the proteasome and autophagy. Indeed, we have shown that Dendra-tau clearance can be accelerated by upregulation of autophagy or by proteasome activation [11, 12]. As described here, the use of proteasome blocking agents (e.g., MG132) and lysosomal acidification inhibitors (e.g., bafilomycin A, chloroquine, or ammonium chloride) allows discrimination between the two clearance pathways and an assessment of the relative contribution of each.

1.1 Overview of the Method

Here, we describe the method we have developed for the photoconversion and analysis of clearance kinetics of Dendra-tagged tau proteins in the neurons of larval zebrafish. However, the protocol is applicable to any Dendra-tagged autophagy substrate and any cell type. An important point to consider is the need for individually labeled cells. This is achieved in our protocol by using a promoter which drives mosaic transgene expression. However, analysis could equally be performed following DNA injections into the fertilized embryo (as described in [10]) to give mosaic expression or by generating transgenic lines with promoters, which label dispersed cell types.

The protocol herein describes the criteria for selection of fish for analysis, the process of embedding, photoconversion, image capture, and the subsequent image analysis. In addition, we include protocols that allow discrimination between proteasomal and lyso-somal/autophagic contributions to substrate clearance.

A brief overview of the protocol is provided here before the detailed Subheadings 2 and 3 below: Fish are sorted at 36 h post-fertilization (h.p.f.) to identify fluorescent Dendra-tau individuals with appropriate mosaicism. At 48 h.p.f., selected fish are anesthe-tized by addition of 3-aminobenzoic acid ethyl ester (MS222, also called tricaine) to the embryo medium and mounted in 0.75% low-melting agarose. Using a confocal microscope, individual neu-rons in the zebrafish spinal cord are photoconverted using a 405 nm (UV) laser targeted at the soma. Digital fluorescent images of red individual photoconverted neurons are acquired immediately after photoconversion and over a further 48 h (at 12 h intervals). Using ImageJ software, the images are then analyzed by measuring the integrated density of the red Dendra signal within a selected region of interest (ROI) drawn around the soma of each photo-converted neuron. The reduction of the fluorescent intensity over time corresponds with the clearance of red protein initially photo-converted and hence to the clearance kinetics of the Dendra-tagged protein.

2 Materials

2.1 Zebrafish Strains

UAS::Dendra-tauWT[cu9], UAS::Dendra-tauA152T[cu10], and EIF1α::Gal4VP16[cu11] zebrafish are maintained on a 14-10 h light-dark cycle, according to standard conditions [13].

2.2 Solutions and Reagents

The following solutions and reagents should be prepared in advance.

1. *Embryo medium*: Embryos and larvae are prepared in embryo medium (EM) containing 5 mM NaCl, 0.17 mM g KCl, 0.33 mM $CaCl_2$, 0.33 mM Mg_2SO_4, 5 mM HEPES at a pH 7.2.

2. *Anesthetic solution*: Prepare aliquots of 4 mg/mL 3-aminobenzoic acid ethyl ester (MS222) in dH_2O. Adjust to pH 7.0, if necessary, using 1 M Tris HCl. Aliquots of MS222 should be stored at −20 °C, and thawed tubes can be kept at

4 °C for up to 1 week. During imaging, fish are anesthetized by the addition of approximately 100 μL 4 mg/mL MS222 into 3 mL EM-containing imaging chamber (final concentration 0.133 mg/mL).

3. *Low-melting point (LMP) agarose:* Prepare 500 μL aliquots of 1.5% LMP agarose in EM and store in the fridge for no more than 6 weeks, or at room temperature for no more than 3 weeks. To melt aliquots of 1.5% agarose, heat the tube at 65 °C using a heat block. During imaging, fish are embedded in 0.75% agarose as described below (Subheading 3.2, **step 2**).

2.3 Equipment

1. *Incubator:* Zebrafish larvae are kept in an incubator at 28.5 °C in the dark during non-imaging periods.

2. *Imaging chambers (confocal dishes):* For imaging, larvae are embedded in agarose in an imaging chamber (Ibidi 60 μ-Dish 35 mm high, glass bottom).

3. *Microscopes:*

   (a) Epifluorescence microscope: Fish must be scored at approximately 36 h.p.f. for appropriate Dendra expression using a fluorescence microscope with EGFP filter. Fish with appropriate mosaic expression of the transgene (strong expression but restricted to a low number of cells, as shown in Fig. 1) will be used for this protocol. Importantly, expression of the transgene needs to be confirmed in the cells of interest (i.e., neurons in the spinal cord). Hence, high magnification is needed to confirm the presence of Dendra-tau positive neurons. We use an Olympus SZX12 stereofluorescence microscope equipped with 12:1 zoom objective and $10\times$ eyepiece magnification to view at $40\times$ magnification.

   (b) Confocal laser scanning microscope: a laser confocal microscope equipped with a $40\times$ objective oil immersion, high numerical aperture (NA) (1.3) objective, excitation/ emission laser and detectors for green Dendra signal (490 nm/507 nm), red Dendra (553 nm/573 nm), and UV laser (405 nm) is utilized. For the photoconversion, a software tool that allows the user to restrict the UV exposure to a single point is needed in order to photoconvert single neurons at specific locations in the soma without affecting the surrounding tissue. This tool is named "bleachpoint" mode in Leica LAS X (version 1.8.1.13759). Lenses with larger NA provide brighter images, but provide a shallower depth of field. An adequate NA is essential to image deep tissue (i.e., spinal cord) in the living fish.

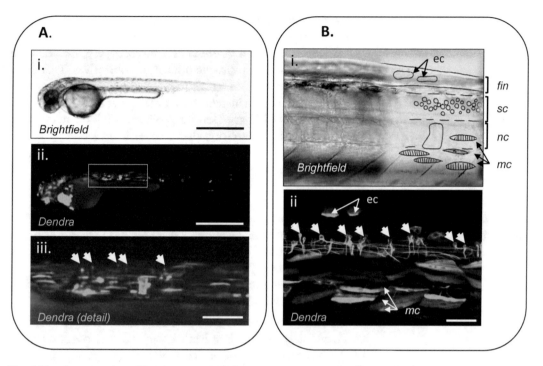

Fig. 1 Mosaic expression of Dendra-tau. **(a)** Epifluorescence microscope: used to identify fish with appropriate levels of expression at 36 h.p.f. before the photoconversion protocol. Sparse mosaic expression is required to visualize individual neurons in the spinal cord rather than clusters of overlapping cells. *(i & ii)* Bright-field and fluorescent images of a fish with appropriate mosaicism at 36 h.p.f. Many cell types can be seen expressing Dendra-tau. Scale bar = 500 μm. Higher magnification (typically 40×) is required to identify Dendra-positive neurons in the spinal cord (white arrows in *iii*). Scale bar = 100 μm. **(b)** Confocal microscope: confocal images obtained using a 40× oil immersion objective. When looking for candidate neurons to photoconvert, bright-field imaging is used first to center and focus on the spinal cord **(sc)** *(i)*. Next, EGFP epifluorescence is used to identify individual neurons (white arrows in *ii*). Many different cell types can be labeled as a consequence of EIF1α-driven expression including epidermal cells **(ec)**, notochord cells **(nc)**, and muscle cells **(mc)**. Scale bar = 50 μm

---

3   Methods

3.1   Zebrafish Crosses and Fish Selection

1. *Zebrafish crosses:* Crosses of UAS::Dendra-tau fish with EIF1α::Gal4VP16 driver fish were performed to produce offspring with ubiquitous but mosaic expression of Dendra-tau, necessary for the visualization of individual neurons.

2. *Collection of the embryos:* Embryos were collected from natural spawning, staged according to established criteria [14] and reared in embryo medium (EM) at 28.5 °C until selection of experimental individuals (*see* **Note 1**).

3. *Selection of fish for the experiment:* The visualization of individual neurons in the spinal cord is crucial for the photoconversion protocol, and hence, previous selection of fish under an

epifluorescent microscope is essential for a successful experiment. At approximately 30 h.p.f., fish are screened for Dendratau expression using a fluorescence microscope equipped with EGFP filters and a high magnification objective. This high magnification is needed to identify individual fish with mosaic expression of Dendra in cells within the spinal cord. Ideally, the selected fish should have Dendra-positive individual neurons in the spinal cord that are not clustered or overlapping (Fig. 1). Selected fish are dechorionated using Dumont no.5 watchmaker's forceps and kept separately in the incubator until use at 48 h.p.f.

3.2 Preparation of Larvae for Imaging

1. *Anesthesia:* Selected larvae with the optimal expression patterns of the transgene are anesthetized at 48 h.p.f. by addition of MS222 to the EM (1 mL of 4 mg/mL MS222 into 30 mL EM, final concentration 0.133 mg/mL). Larvae need to be anesthetized to prevent them from moving while embedding and to ensure the correct orientation.

2. *Embedding:* Larvae are next embedded into LMP agarose in a confocal imaging chamber. Melt 1.5% LMP agarose aliquots to 65 °C (prepare one aliquot per experimental group) (*see* **Note 2**). Prior to adding the larvae to a single aliquot, remove one tube of agarose from the heat block and invert several times to cool. Collect 10–15 anesthetized larvae for each experimental condition in a plastic Pasteur pipette within a volume of 500 μL EM (containing MS222). Pipette the 500 μL containing the fish into the 500 μL aliquot of warm, melt 1.5% LMP agarose, and mix gently by pipetting up and down. This creates a final mixture of 0.75% LMP agarose containing anesthetic and fish. Pipette the 1 mL 0.75% agarose containing the fish into the confocal imaging chamber. Using a dissecting microscope, orientate the fish on their side with forceps as close as possible to the glass bottom, as this will minimize the focal distance to the lens when imaged. This needs to be done quickly while the agarose is cooling but before it sets. This step requires some practice. Organizing the fish in columns (*see* Fig. 2) facilitates moving from one fish to the next one when imaging. Set the dish aside and avoid touching it until agarose is set (*see* **Note 3**). Once agarose is set, imaging chambers can be topped up with EM without anesthetic (*see* **Note 4**).

3.3 Confocal Imaging

The protocol explained below is specific for a confocal Leica TCS SP8 running LAS X (version 1.8.1.13759) software. Any inverted confocal microscope with lasers appropriated for green/red Dendra and UV wavelengths is suitable for this method. Although not essential, using a temperature-controlled imaging chamber at 28.5 °C is recommended.

A.

B.

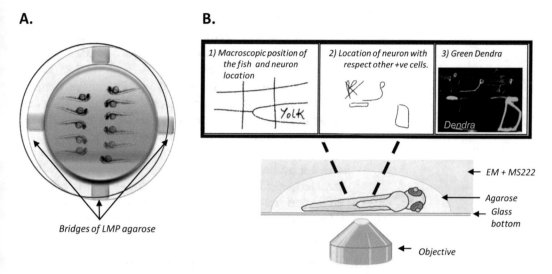

Bridges of LMP agarose

Fig. 2 Preparation of fish for imaging. (**a**) Embedding: fish are embedded and orientated on their side in 0.75% LMP agarose in a 35 mm imaging chamber. To avoid the detachment of the agarose from the bottom, additional agarose can be added to the walls of the dish to anchor the central agarose in position. These are labeled as "*bridges of LMP agarose*." The parallel positioning/orientation of the fish helps when moving from one fish to the next one while imaging. (**b**) Visualization and schematic representation of neurons: the exact location of photoconverted neurons is essential in order to find the same neuron at the subsequent imaging time points. An easy way to record the position of the neuron is to make a schematic drawing. Once a neuron has been selected using conventional epifluorescence filters for 488 nm excitation, position the neuron(s) in the center of the field of view and switch to bright-field. Draw a schematic diagram of the orientation of the fish and large structures in the field of view such as the yolk or head (1). In this example, the fish is orientated with the head to the right and the yolk to the bottom of the field of view and the neuron is to the right of the yolk extension. Switch back to epifluorescence to draw the shape of the neuron(s) and its location with respect other Dendra-positive cells such as notochord cells, muscle cells, or other neurons (2). Your sketch should match the green Dendra image (3)

Dendra protein can be photoconverted with 488 nm light as well as 405 nm light. This range covers excitation used to image the green signal (i.e., 488 nm laser or a standard GFP excitation filter of appropriate characteristics) [15]. For this reason, the exposure to 488 nm light should be minimized by reducing the intensity of the epifluorescence during the selection of the fish and localization of the neurons and also during the imaging process. In addition, no images of the green signal are acquired until after first time point ($T = 0$).

3.3.1 Preparation for Live Imaging

1. Switch on confocal microscope and lasers according to manufacturer's guidance and then open the imaging software.

2. Change objectives to a 40× oil-immersion lens and place the confocal imaging chamber with fish embedded in agarose and EM onto the stage while illuminating it with bright-field (BF) light. Add 100 μL of MS222 4 mg/mL into the 3 mL EM (MS222 final concentration of 0.133 mg/mL) to avoid fish movements (*see* **Note 5**).

3. Focus on the spinal cord area of the fish.

4. Set up your imaging program with the following settings:

   (a) Image acquisition and format: images are 1024×1024 pixels acquired at 12-bit depth. A zoom factor 3.00 is applied in order to identify and visualize single neurons and to mark off the photoconversion point in the soma more accurately, in the center of the conical part of the cell body from which the axon emerges and where greatest green fluorescence signal can be seen (Fig. 3). Do not select a point within the nucleus as less protein will be targeted, and DNA UV exposure could have detrimental effects on the cell. These settings result in a pixel size of 94 nm and a $96.88\ \mu m \times 96.88\ \mu m$ field of view.

   (b) Acquisition mode: *xyz* for z-stacks.

   (c) Dwell time: 0.4 μs (laser speed of 600 Hz).

   (d) Bidirectional mode: ON. A bidirectional mode accelerates the acquisition time.

   (e) Line average: 2 (this will improve the quality of the image).

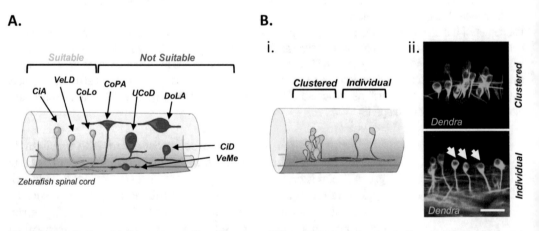

Fig. 3 Fish selection. (**a**) Neuron types: there are many different types of neurons in the zebrafish spinal cord. To minimize the variability in the assay, utilize the same neuron type. As no markers for neuronal type are used, the selection criteria are based on the cell shape and axon trajectories. This assay was initially published using **CiA** (homologue of mammalian V1 spinal interneurons), **VeLD** (homologue to mammalian V2b), and **CoLo** (inhibitory interneurons) neuronal types. We have not used neurons such as **CoPA** (commissural primary ascending interneuron), **UCoD** and **VeMe** (homologues to mammalian V3 interneurons), **DoLA** (dorsal longitudinal ascending interneurons), or **CiD** (homologue to V2a neurons). (**b**) Clustered versus individual neurons: for the photoconversion and for measuring the reduction in the red Dendra-tau intensity over time, the somas of selected neurons must not overlap with other neurons, hence, clustered neurons are not suitable for this method. Individual and well-separated neurons such as those represented in (i) or imaged in (ii, lower panel, white arrows) are easy to identify and facilitate the positioning of the "bleachpoint" tool to photoconvert the Dendra-tau within a single neuron without affecting surrounding Dendra-positive cells. Scale bar = 30 μm

(f) Lasers (*see* **Note 6**):

- Diode 405 nm (UV light): needed for the photoconversion and used at 10% power (equivalent to 250 µW).

- Argon 488 nm: used to determine the z-stack and image green Dendra immediately after photoconversion. In order to avoid bleaching or photoconversion under blue light, laser power is reduced to 5% (equivalent to 150 µW).

- DPSS 561 nm: used to image Dendra-tau red signal. Some optimization might be necessary in order to collect the maximal red Dendra signal after photoconversion, which directly depends on the level of expression of the transgene (and hence, levels of green protein to be photoconverted), and 561 nm laser power. 7% laser power was used to collect red signal in our published studies (equivalent to 118 µW) [11].

(g) Sequential acquisition mode: the spectra from both green and red Dendra signal overlap [16], therefore, green and red Dendra-tau signals are scanned sequentially to avoid cross talk.

- Sequence 1—Green: used for excitation at 488 nm, with 5% of laser power to avoid spontaneous photoconversion. PMT-type detector collects emitted green light from 511 to 564 nm to avoid overlap with red signal emission spectrum. Gain values around 700–850 V are sufficient to visualize the green signal (*see* **Note 7**).

- Sequence 2—Red: used to image red signal by exciting at 561 nm and to collect red emitted light between 574 and 674 nm using a hybrid detector HyD (gain set at 100%). Contrary to PTM detector used for green signal, HyD detectors are more sensitive, and only 7% of the power laser is used to excite the sample.

3.3.2 Photoconversion and Live Imaging of Initial Time Point (T = 0)

1. Once the spinal cord is in focus using the BF settings, switch to 488 nm epifluorescence light to find the individual neurons to photoconvert. Keep the intensity of 488 nm emission light low to avoid spontaneous photoconversion of Dendra. *See* **Note 8** regarding how to identify suitable neurons for the assay.

2. Once suitable neurons are selected under the GFP epifluorescence filter, center the neurons in the field of view and go back to BF to make a schematic drawing of the position of those neurons in the fish with respect to macroscopic structures, such as yolk extension, somite, or head. Also, draw the schematic position of the neuron(s) relative to other Dendra-positive cells you might see in the surroundings. These drawings are used to

map the neurons and will help you identifying the same neuron at the successive time points (*see* Fig. 3). Close the shutter during drawing to avoid bleaching/photoconversion. The sketch should only take about 5 s to draw.

3. Using the preview option in the confocal imaging software and visualizing using the green channel, set a *z*-stack for the neuron or group of neurons covering the whole thickness of the somas (no red signal should be seen at this stage):
   - With the *z*-position knob define "begin" and "end" of the *z*-stack in "Live" mode. Set the number of steps such that it is equal to the number of microns (µm) in the z-stack (i.e. each step is 1 µm). Only one image per micron will be acquired. This parameter has to be kept the same for all neurons and for the subsequent acquisition time points. *Z*-stacks for single neurons vary between 9 and 13 µm.

   - Center the image in the middle point of the neuron of interest by clicking the center button in the *z*-stack panel or manually and stop the live preview. An image of the center of the neuron remains on the screen. In case of multiple neurons at different planes, a center image of each neuron is needed before selecting the photoconversion point.

4. (a) Photoconversion of individual neurons using the bleach-point (or similar) tool (e.g., Leica SP8 with bleachpoint tool).
   - Select the bleachpoint tool and set 3 s of exposure in the "bleachpoint panel."

   - Turn on UV 405 nm laser to 10% and reduce to 0% the power of the other lasers (i.e., turn the 488 nm laser down from 5 to 0% in case you are in sequence 1).

   - Select photoconversion points by clicking in the soma of the neuron right in the center of the conical shape from which the axon projects (*see* **Note 9**).

   - Click start to initiate photoconversion (all neurons will be photoconverted sequentially, taking 3 s per selected point).
     (b) For confocal microscopes other than Leica SP8:
     Select a tool that immobilizes the laser to a specific point and restricts the exposure to a certain point of the sample (we have used the cell body cone of the neuron in our studies). Set UV 405 nm exposure to 10% of the laser (equivalent to 250 µW) and a duration of the exposure of 3 s for each point/neuron selected.

5. Immediately after photoconversion, images of the green and the red signal should be acquired sequentially. A green image is only captured at the initial photoconversion time point (not subsequent ones) and serves as a reference image of the entire

field of Dendra expression, as only photoconverted neurons can be seen in the red channel. The reduction of the green signal in photoconverted neurons, relative to neurons which have not been photoconverted, also indicates how successful the photoconversion was (*see* **Note 10**).

6. Look for other neurons in the same fish or move to next larva and repeat photoconversion (*see* **Note 11**).

7. Once photoconversion is complete for an experimental group, remove the imaging chamber from the stage and replace the EM immediately with fresh EM without anesthetic. If compounds are being used in the experiment, these should also be added at this ($T = 0$) point (*see* Subheading 3.5).

8. At the end of the photoconversion protocol, confocal imaging chambers with embedded larvae and fresh medium are returned to the incubator and kept in the dark at 28.5 °C until the next imaging time point. Fish must always be kept in the incubator during non-imaging intervals.

3.3.3 Live Imaging at 12, 24, 36, and 48 h Time Points

For subsequent imaging at 12-h intervals, epifluorescence GFP is used to identify the photoconverted neurons and to position them for imaging. Confocal images will only be captured for the red channel using the same settings as Sequence 2 at $T = 0$.

1. Add anesthetic to the EM (100 μL of 4 mg/mL MS222 into the 3 mL EM to a final concentration of 0.133 mg/mL) and place the imaging chamber onto the microscope stage (*see* Subheading 3.2).

2. Check that software acquisition settings remain the same as during photoconversion protocol (including *xyz* mode, size, depth, speed, zoom, and line average). Otherwise, change accordingly.

3. Only the DPSS 561 nm light source is used to image the photoconverted red signal during these time points, so the sequential mode is no longer needed. The excitation/emission values when using this laser must correlate with the values set in Sequence 2 in the sequential mode used during the photoconversion at the first time point (*see* Subheading 3.3.1). All settings must be kept the same to subsequently compare the change in red intensity in captured images.

4. Localize photoconverted neurons using schematic drawings, BF, and epifluorescence GFP (minimize the use of GFP fluorescence where possible) (*see* **Note 12**).

5. Using live imaging or preview of the red signal, set a *z*-stack for the photoconverted neuron/neurons using the red channel (similarly to **step 3** in Subheading 3.3.2). *Z*-stack step size must be set at 1 μm.

6. Once all neurons from an experimental group are imaged, replace medium with new EM without anesthetic and return the dishes to the incubator at 28.5 °C (*see* **Note 13**).

3.4 Analysis of Images

Confocal digital images are analyzed using FIJI (ImageJ) software. Only images of the red Dendra-tau signal are used in the measurement.

3.4.1 Z-Stack Projection

1. Open files in FIJI.
2. Create a hyperstack of the z-stack using the maximum intensity projection settings.
3. Save the maximum intensity projection image as a tiff file.
4. Repeat **steps 1–3** for all z-stacks. Care must be taken to generate unique file names for each stack. For example, following the rule (Treatment)_ (Fish).(Neuron)_ (time after photoconversion), i.e., Unt_2.3_0 h for an image of the third neuron photoconverted taken in the second fish in the untreated group, immediately after photoconversion.

3.4.2 Quantification of the Red Signal

1. Open all tiff files of images of the same neuron taken at the different time points (i.e., 0, 12, 24, and 36 h post-photoconversion) (*see* **Note 14**).
2. Draw a ROI around the soma of the neuron immediately after photoconversion ($T = 0$) using the polygonal tool. In the rest of the images, maintain the ROI shape as much as possible in order to select similar areas to the original image (*see* **Note 15**).
3. Measure the area, mean, and integrated density of the ROI.
4. Copy the values into an excel sheet.

3.4.3 Analysis of the Red Signal Values

1. The integrated density corresponds with the measurement of the mean intensity of the red signal within a defined area (the ROI). Set the value of the integrated density at $T = 0$ (immediately after photoconversion) to 100% for each neuron.
2. Express the integrated density for each time point relative to the 100% value at $T = 0$ (for example, if the IntDen0hr $= 900,000$ at $T = 0$, this is set at 100%. Therefore, IntDen12hr $= 450,000$ at 12 h corresponds to 50%).
3. Calculate the mean percentage, standard deviation, and the standard error of the mean for each time point. There should be approximately 50–60 values per time point, corresponding to 50–60 neurons from 12 to 15 fish.
4. Represent the changes in the intensity of the red signal in a line chart using percentage of initial red intensity on the y-axis and time on the x-axis (Fig. 3).

3.5 Additional Protocols

3.5.1 Assay for Lysosomal Protein Clearance

Protein degradation under the acidic and proteolytic conditions within the lysosomal compartment is one of the major pathways for protein clearance in cells. Macroautophagy is the major contributor to lysosomal protein degradation. However, the lysosome is the final stage of other clearance pathways, such as microautophagy, chaperone-mediated autophagy, and direct protein import pathways [17]. Treatment with lysosomal inhibitors abrogates all these clearance systems and hence leads to slower degradation of proteins cleared via these pathways. Ammonium chloride (NH_4Cl) is an inhibitor of lysosomal acidification and, therefore, inhibits the final steps of all autophagic pathways. To assess lysosome-dependent protein clearance (and hence autophagy-dependent clearance), 10 µM NH_4Cl can be added immediately after photoconversion into the EM (using a 10 mM NH_4Cl stock in distilled water) and at each EM change. Imaging of red signal at 12 h intervals showed a significant reduction in the clearance rate of Dendra-tau, in both wild-type and mutant A152T forms, upon NH_4Cl treatment [11], with changes observed as early as 12 h post-photoconversion.

3.5.2 Assay for Proteasomal Protein Clearance

A second major protein degradation system is the proteasome, a barrel-shape protein complex that degrades ubiquitinated proteins. MG132 is a pharmacological blocker of the proteolytic activity of the 26S proteasome complex and hence inhibits proteasomal degradation of cytosolic proteins. Treatment with 100 µM MG132 (using a 100 mM MG132 stock in DMSO) in the EM at $T = 0$ and replenished daily blocks the proteasome and results in slower Dendra-tau protein clearance for wild-type tau. Treatment of mutant Dendra-tau A152T with 100 µM MG132, however, did not modify red intensity decline compared to untreated fish, and subsequent validation experiments demonstrated that proteasome clearance is impaired in A152T-tau fish and, therefore, cannot be further slowed by pharmacological inhibition [11].

3.5.3 Investigation of Compounds for Their Effects on Protein Clearance

To evaluate the effects of drug treatment on the clearance rate of Dendra-tau, compounds at the appropriate concentration are added immediately after photoconversion and replenished at least daily. We have evaluated the effects of the autophagy up-regulators rapamycin, clonidine, and rilmenidine on Dendra-tau clearance kinetics. Once photoconversion is complete, EM with MS222 is removed and replaced by fresh EM (without anesthetic) containing the appropriate concentration of the compound (final concentrations are 30 µM rapamycin, 30 µM clonidine, or 50 µM rilmenidine). Drugs must be replenished every day (ideally, every time the medium is changed) to ensure the concentration is maintained and to facilitate diffusion of the compound through the agarose. Induction of autophagy by the treatment with up-regulators enhances Dendra-tau clearance, resulting in a downward shift in the clearance curve representing the percentage of initial red intensity (*see* Fig. 4).

A.

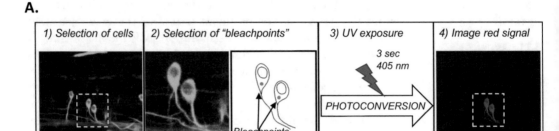

B.

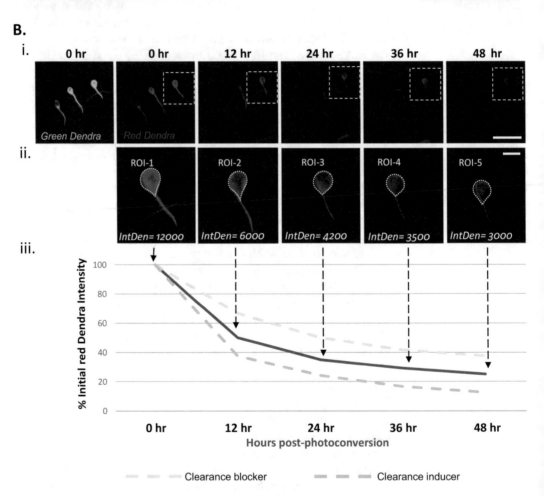

- - - Clearance blocker   - - - Clearance inducer

Fig. 4 Imaging and analysis. (**a**) Setting points for photoconversion: Once the neurons have been selected (1), green Dendra-tau cells are photoconverted by exposure to 405 nm (UV) light. The photoconversion occurs by positioning the UV laser on a specific point (here termed "bleachpoint") within the soma of the selected neuron and exposing it for 3 s. The *bleachpoint* must be allocated in the center of the conical shape of the soma (2) from which the axon emerges (purple dots); here, the intensity of the green signal is often stronger than in the rest of the soma. The location of this "bleachpoint" will determine the success of the photoconversion. For example, a more central position within the soma usually corresponds to nuclear localization and the lowest Dendra signal. After UV exposure (3), green Dendra is converted into red Dendra, which rapidly diffuses along

3.6   Results

The measurement of the red signal integrated density reflects the amount of photoconverted Dendra-tau present at each time point. The reduction in the integrated density (intensity) of red Dendra-tau over time corresponds to the clearance kinetics of the red Dendra-tau protein. It is important to note that Dendra-tau protein is continuously synthesized, and this is evident as green Dendra-tau. The green fluorescence is greatly reduced by photoconversion. Over time, the green fluorescence will increase as a result of the continuous transcription and translation of the transgene, whereas red fluorescence, unaffected by expression levels, reduces over time due to tau clearance. The relative value of red intensity compared to the initial red intensity can be interpreted as the clearance rate of the Dendra-tagged protein (*see* Fig. 3).

The compound concentrations and imaging intervals described here are sufficient to evaluate the effect of different autophagy and proteasome-related compounds on Dendra-tau clearance over a period of 48 h. The use of other compounds would require optimization of the drug concentration, imaging intervals, and length of the experiment in order to obtain significant changes compared to the untreated group.

3.7   Conclusions and Perspectives

Measurement of the clearance rate of Dendra-tau (and potentially other Dendra-tagged proteins) using this method allows evaluation of pharmacological and genetic modifiers of protein clearance such as modulators of autophagy or proteasome function. Hence, the treatment with different compounds that affect these processes will alter the clearance rate of Dendra-tagged proteins and can be used to analyze the integrity of the different protein clearance pathways.

Typically, compound treatments are administered immediately after photoconversion and refreshed daily. For example, we have found that treatment with autophagy blockers such as ammonium chloride slows the clearance of both wild-type and A152T forms of Dendra-tau (an autophagy substrate). However, while proteasomal inhibition by addition of MG132 slows the clearance of the wild-type form, it has no effect on the clearance rate of the tau variant A152T [11]. These results suggest that either Dendra-tau A152T protein cannot be degraded by the proteasomal pathway or that the

Fig. 4 (continued) the axon (4). Scale bar = 50 μm. (**b**) Time-course imaging and analysis: The pool of photoconverted red Dendra-tau is imaged immediately after photoconversion (0 h, $T=0$) and at 12-h intervals for the following 48 h (i) (Scale bar = 50 μm). The clearance of the red signal is measured by selecting regions of interest (ROI) around the somas of red neurons (ii) and measuring the integrated density (*IntDen*) of these ROIs at the different time points (Scale bar = 10 μm). The reduction in the percentage of initial red Dendra intensity over time represents the clearance kinetics of the Dendra-tagged protein (iii). The use of up or downregulators of clearance pathways affects the clearance kinetics of Dendra-tau; an upward shift in the curve is observed with blockers that slow the clearance of Dendra-tau, a downward shift is observed when using inducers that accelerate clearance pathways, such as autophagy and proteasomal degradation

proteasome is not functional in these fish, and additional experiments were performed to confirm the latter [11]. Similarly, intrinsic variations within a protein, such as different posttranslational modifications, can also affect its degradation rate, not by affecting the function of the degradation systems but by changing the availability/susceptibility of the protein to be cleared. In the initial experimental design, it is important to consider the protein half-life.

This protocol can also be employed to assess genetic modifiers, which may be involved in protein clearance pathways. Indeed, one of the benefits of the zebrafish model is the ease of genetic manipulation [18, 19]. However, transient genetic modulation is typically performed by injection of the over-expression or knockdown construct into the 1-cell-stage embryo; therefore, expression of the gene of interest would be affected throughout development, not only at photoconversion and the subsequent period. In such scenarios, it is necessary to ensure that the levels of green Dendra-tagged protein available to photoconvert are comparable in control and experimental groups before the photoconversion. This problem may be circumvented with tools allowing inducible expression [20–22].

A common characteristic of neurodegenerative diseases is the accumulation and aggregation of proteins into insoluble inclusions. The assay described here is likely to be measuring soluble tau since insoluble tau was only detected at 6 d.p.f. in our model, whereas this assay is performed from 2 to 4 d.p.f [11]. One can presume that insoluble aggregates have different clearance kinetics, and hence, a different photoconversion protocol, imaging intervals, and assay duration may be needed for their analysis. In addition to autophagy and proteasomal clearance, this model may prove to be a suitable tool for assessing protein secretion and spreading of aggregate-prone proteins [23, 24] although further validation would be necessary to determine whether such clearance routes exist in any given model.

4   Notes

1. Addition of 1-phenyl-2-thiourea (PTU) to a final concentration of 0.003% in EM will block pigment formation and will facilitate imaging.

2. Aliquots of 1.5% LMP can be left in the heat block overnight to save time on the day of the experiment.

3. Once the agarose containing the fish has set, additional 1.5% agarose can be added on top to adhere to the walls of the imaging dish (see Fig. 2).

4. Do not add EM to agarose when transporting the imaging chambers. Movement of the EM can cause detachment of agarose from the glass bottom and will free the embedded fish. Remove EM from the dish when it is taken out from the incubator and add it when you have the dishes next to confocal microscope. Keep the dish with embedded fish dry ONLY during transporting (no more than 2–3 min to avoid agarose dry out).

5. It is essential that fish remain anesthetized for the duration of the experiment.

6. Optimization is required for each confocal microscope and for the different excitation light sources. This protocol details the values used in the original method described, including the power of the lasers employed. The power of the lasers varies with each laser source and hence will require optimization.

7. A normal detector PMT is used to image the green signal as green image is not used for quantitative purpose.

8. How to identify suitable neurons for the assay:
   - There are many different types of neurons in the spinal cord. Ideally, neurons used should be of the same type, however, only shape can be used to score these neurons as no other fluorescent markers can be used in the experiment. Loco-motor interneurons VeLD type (homologue toV2b) are ideal for this experiment as they are abundant and have a characteristic and easily recognizable shape (*see* Fig. 4).

   - Multiple neurons can be captured in a single image, however, overlapping neurons are not suitable as it will not be possible to quantify the red signal from individual cells.

   - A large number of neurons is needed for this assay as some of the photoconverted neurons might either die or divide, and hence are not suitable for analysis in the experiment. Importantly, neurons, as living cells, might change their shape too. Drastic changes in shape that affect the measured area cannot be used either. A fluctuation of 20–25% of the area is acceptable. Larger changes would affect the intensity measurement in a non-clearance-dependent way.

9. In the case of multiple neurons in the same image, find the middle point of each cell, using live imaging settings, before selecting the bleachpoint. Previously selected points will be saved while imaging.

10. Some green signal remains after photoconversion, suggesting that not all Dendra-tau protein is photoconverted. We have not achieved complete green to red conversion of a neuron without either killing the cell or bleaching completely any fluorescent signal. However, the remaining green Dendra is useful to

localize the neurons at the time points following photoconversion.

11. The number of neurons used in this assay depends on the type of experiment and parameters analyzed. A minimum of 50–60 neurons per group (typically 12–15 fish) was needed to obtain statistically significant differences in the original paper for this method [11].

12. Multi-position recordings can be implemented by coupling a stepping motor to the stage of the confocal microscope. However, because of the growth of the fish and intrinsic movement of neuronal soma in the spinal cord, positions may change and require some adjustments.

13. To improve fish survival, medium can be replaced every 2–3 h. Also, air can be bubbled into the medium carefully with a Pasteur pipette to improve oxygenation of medium and oxygen diffusion through agarose.

14. The visualization of all images acquired from the same neuron on the computer screen helps to identify and discard those neurons that are not suitable for the experiment because of a drastic change in the shape, cell divisions, or cell death. Moreover, the reduction in the red Dendra intensity might be also seen by eye.

15. The shape of some neurons may change substantially, and hence, these will not be suitable for analysis. As the larva grows, neurons may rotate and their axons elongate, whereas the shape of the soma does not change substantially. However, on occasion, the soma expands or shrinks affecting the intensity of that neuron as a consequence of a drastic change in the volume rather than changes in the levels of the protein. The calculation of the integrated density is based on the area and the mean intensity of a ROI around the soma, and hence, any cell bodies that display drastic changes in the area must be excluded from analysis.

Acknowledgments

We are grateful to the UK Dementia Research Institute at the University of Cambridge (funded by the MRC, Alzheimer's Research UK and the Alzheimer's Society) (DCR), The Tau Consortium, and Alzheimer's Research UK for funding. We are grateful to Dr. Kevin O'Holleran, Cambridge Advanced Imaging Centre, for technical help and advice in establishing this protocol and for proofreading the manuscript.

References

1. Lopez A, Fleming A, Rubinsztein DC (2018) Seeing is believing: methods to monitor vertebrate autophagy in vivo. Open Biol 8 (10):180106

2. Das AT, Tenenbaum L, Berkhout B (2016) Tet-On systems for doxycycline-inducible gene expression. Curr Gene Ther 16 (3):156–167

3. Santacruz K et al (2005) Science 309 (5733):476–481. Tau suppression in a neurodegenerative mouse model improves memory function.

4. Chudakov DM, Lukyanov S, Lukyanov KA (2007) BioTechniques 42(5):553, 555, 557 passim. Using photoactivatable fluorescent protein Dendra2 to track protein movement.

5. Berger Z et al (2006) Hum Mol Genet 15 (3):433–442. Rapamycin alleviates toxicity of different aggregate-prone proteins.

6. Ravikumar B, Duden R, Rubinsztein DC (2002) Hum Mol Genet 11(9):1107–1117. Aggregate-prone proteins with polyglutamine and polyalanine expansions are degraded by autophagy.

7. Ravikumar B et al (2004) Nat Genet 36 (6):585–595. Inhibition of mTOR induces autophagy and reduces toxicity of polyglutamine expansions in fly and mouse models of Huntington disease.

8. Webb JL et al (2003) J Biol Chem 278 (27):25009–25013. Alpha-Synuclein is degraded by both autophagy and the proteasome.

9. Tsvetkov AS et al (2013) Nat Chem Biol 9 (9):586–592. Proteostasis of polyglutamine varies among neurons and predicts neurodegeneration.

10. Moreau K et al (2014) Nat Commun 5:4998. PICALM modulates autophagy activity and tau accumulation.

11. Lopez A et al (2017) A152T tau allele causes neurodegeneration that can be ameliorated in a zebrafish model by autophagy induction. Brain 140

12. VerPlank JJS et al (2020) cGMP via PKG activates 26S proteasomes and enhances degradation of proteins, including ones that cause neurodegenerative diseases. Proceedings of the National Academy of Sciences 117 (25):14220–14230

13. Westerfield M (2007) The zebrafish book. A guide for the laboratory use of zebrafish (Danio rerio), 5th edn. University of Oregon Press, Eugene

14. Kimmel CB et al (1995) Dev Dyn 203 (3):253–310. Stages of embryonic development of the zebrafish.

15. Chudakov DM, Lukyanov S, Lukyanov KA (2007) Nat Protoc 2(8):2024–2032. Tracking intracellular protein movements using photoswitchable fluorescent proteins PS-CFP2 and Dendra2.

16. Adam V et al (2009) Biochemistry 48 (22):4905–4915. Structural basis of enhanced photoconversion yield in green fluorescent protein-like protein Dendra2.

17. Tekirdag K, Cuervo AM (2018) J Biol Chem 293(15):5414–5424. Chaperone-mediated autophagy and endosomal microautophagy: Joint by a chaperone.

18. Liu J et al (2017) Hum Genet 136 (1):1–12. CRISPR/Cas9 in zebrafish: an efficient combination for human genetic diseases modeling.

19. Sassen WA, Köster RW (2015) Adv Genomics Genet 5:151–163. A molecular toolbox for genetic manipulation of zebrafish.

20. Gerety SS et al (2013) Development 140 (10):2235–2243. An inducible transgene expression system for zebrafish and chick.

21. Tallafuss A et al (2012) Development 139 (9):1691–1699. Turning gene function ON and OFF using sense and antisense photomorpholinos in zebrafish.

22. Zhou W, Deiters A (2016) Angew Chem Int Ed Engl 55(18):5394–5399. Conditional Control of CRISPR/Cas9 Function.

23. Lim J, Yue Z (2015) Dev Cell 32 (4):491–501. Neuronal aggregates: formation, clearance, and spreading.

24. Goedert M, Eisenberg DS, Crowther RA (2017) Annu Rev Neurosci 40:189–210. Propagation of tau aggregates and neurodegeneration.

Chapter 8

Methods for Studying Axonal Autophagosome Dynamics in Adult Dorsal Root Ganglion Neurons

Xiu-Tang Cheng, Kelly A. Chamberlain, and Zu-Hang Sheng

Abstract

Neuronal autophagy is an important degradative pathway for maintaining cellular homeostasis essential for neuronal survival and functions. In neurons, mature acidic lysosomes are relatively enriched in the soma. Long-distance transport of autophagosomes generated at distal axons is a key step for efficient degradation of autophagic vacuoles (AVs) via lysosomes in the soma. However, the mechanisms underlying the regulation of axonal autophagosome trafficking remain largely elusive. Here, we provide several methodologies for studying axonal autophagy, including primary culture of adult dorsal root ganglion (DRG) neurons, live-cell time-lapse imaging, and quantitative analysis of autophagic dynamics. DRG neurons can be prepared from adult mouse and rat tissue, making it feasible to study axonal autophagic events in neurons that better resemble in vivo conditions characteristic of healthy and disease states. Using these methods, we established a "motor-sharing" mechanism that drives autophagosome transport toward the soma. This trafficking route is crucial for neurons to maintain effective autophagic clearance through lysosomal degradation in the soma. The methods described in this chapter provide practical guidelines for the further study of neuronal autophagy in various physiological and pathological circumstances.

Keywords Dorsal root ganglion (DRG) neuron, Live-cell time-lapse imaging, Confocal microscopy, Axonal transport, Autophagosome

1 Introduction

Macroautophagy (hereafter referred to as "autophagy") is an essential catabolic pathway for neuron growth, development, synaptic functions, and survival. It maintains cellular homeostasis by degrading protein aggregates and damaged organelles, thereby recycling basic metabolites for protein synthesis [1, 2]. Autophagy is composed of the following principal stages: (1) formation of an isolation membrane (or phagophore) with the aid of multiple signaling and protein modification assemblies; (2) elongation of the isolation membrane surrounding the substrates to be degraded and gradual closure of the two tips to form a nascent double-membrane autophagosome; (3) fusion of the nascent autophagosomes with lysosomes or late endosomes, forming hybrid organelles called

Ben Loos and Esther Wong (eds.), *Imaging and Quantifying Neuronal Autophagy*, Neuromethods, vol. 171,
https://doi.org/10.1007/978-1-0716-1589-8_8, © Springer Science+Business Media, LLC, part of Springer Nature 2022

autolysosomes or amphisomes, respectively; (4) lysosomal degradation, in which the internal contents of autophagosomes are digested by active lysosomal enzymes, yielding basic metabolites that are released into the cytoplasm and recycled for protein synthesis [3–6]. Since neurons are postmitotic cells that cannot dilute cellular waste through division, the autophagic degradation pathway is critical for neuronal health. This is evidenced by several rigorous studies showing that depletion of autophagy-related proteins is associated with neurodegeneration [7–11]. Specifically, defects in autophagy functions are closely linked to various neurodegenerative disorders, such as Alzheimer's disease, Parkinson's disease, and amyotrophic lateral sclerosis (ALS) [12–15]. Clearance of aggregated proteins by elevating autophagy functions can help ameliorate the pathological phenotypes in disease-linked mouse neurons [16, 17]. However, the precise mechanisms regulating autophagosomes in neurons are just beginning to be revealed.

One unique feature of neurons is their extremely polarized structure. Usually, neurons have a relatively small cell body, complex dendritic arbors, and an especially long and extended axon with axonal terminal branches. In humans, peripheral nerves and corticospinal tracts give rise to axons up to 1 m in length; individual neurons of the substantia nigra pars compacta give rise to 4.5 m of axon, once all the branches have been summed [18, 19]. Since cellular machineries for DNA replication, mRNA, and protein synthesis, as well as degradative organelles, are predominantly localized in the soma, neurons face the exceptional challenge of delivering autophagosomes generated at the distal axons to the cell body for complete degradation [20–23]. Many neurodegenerative disorders are accompanied by axonal dystrophy, characterized by bulbous swellings consisting of accumulated organelles and degradative material associated with the autophagy-lysosomal pathway [24]. Such defects suggest impaired axonal transport of autophagosomes and/or reduced lysosome degradation capacity. However, the mechanisms underlying these pathologic changes remain obscure. Therefore, investigation into the regulation of axonal autophagosome transport is an important topic of consideration for researchers working in the neurodegeneration field at large.

In the current protocol, we utilize adult mouse dorsal root ganglion (DRG) neurons as our cell model. DRG neurons are advantageous for studying the regulation of axonal autophagosome transport for the following reasons. First, all neurites of DRG neurons are tau-positive axons (Fig. 1), saving the researcher time spent searching for individual axons among matrices of dendrites in prevalent cortical or hippocampal neuron culture systems. Second,

Tau   MAP2   Merged

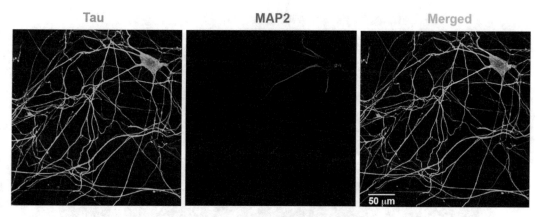

Fig. 1 All neurites of a DRG neuron are axons. Immunostaining images showing in vitro condition, a DRG neuron usually has multiple neurites, and all the neurites are tau-positive (green) axons. Only the very beginning segment of the neurites has a light Map2 staining. DIV3 adult mouse DRG neurons are immunostained with antibodies against tau, the axon marker and Map2, the dendrite marker. Scale bar: 50 μm

dissociated DRG neurons can be cultured from adult animals, offering the possibility to study mature neurons better resembling in vivo characteristics [25]. Third, the onset of pathology in many models of neurodegeneration occurs well into adulthood, when isolating cortical neurons from adult animals becomes very difficult or technically infeasible. With the current protocol, DRG neurons can be isolated and cultured from mice aged between P21 and 2 years old, providing an excellent opportunity for mechanistic studies to be linked to the age-related disease phenotypes. Fourth, DRG neurons belong to the peripheral nervous system and thus have stronger regrowth capacity and a higher survival rate in vitro. This likely contributes to the fact that DRG neurons often have more robust autophagic events than cortical neurons, as our empirical experience has demonstrated that autophagy can only be induced in relatively healthy neurons.

Using the DRG model, we provide a detailed procedure for live-cell time-lapse imaging using confocal microscopy and quantitative analysis of autophagic dynamics via ImageJ. Using GFP-LC3 as an autophagosome marker, we have shown that [1] a large pool of axonal autophagosomes is generated at the axon terminal; [2] nascent autophagosomes undergo fusion with late endosomes and thus share their motor proteins to achieve motility; and [3] after fusion, the hybrid organelles present a dominant retrograde motility back to the soma (Fig. 2) [23, 26]. The protocols described in this chapter should be helpful for researchers studying neuronal autophagy dynamics in various physiological and pathological circumstances.

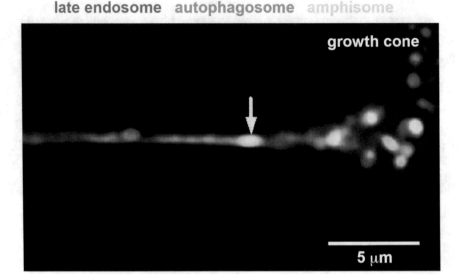

Fig. 2 Dynamic de novo autophagosome biogenesis, fusion, and retrograde transport in a DRG neuron growth cone. Autophagosome biogenesis is very robust in DRG neuron growth cones. Red puncta are Rab7-labeled late endosomes. Green puncta indicate nascent autophagosomes. The yellow arrow denotes an amphisome, a hybrid organelle generated by fusion of late endosomes and autophagosomes in axons. DRG neurons were transfected with GFP-LC3 and mRFP-Rab7 at DIV0 and imaged at DIV3 after incubation with serum-free medium for 3 h. Live-cell image acquisition was performed at the growth cone of the axon. Scale bar: 5 μm

2   Materials

2.1   Primary Mouse DRG Neuron Culture

2.1.1   Medias, Solutions, and Reagents

1. Artificial cerebrospinal fluid (ACSF): 125 mM NaCl, 25 mM NaHCO$_3$, 2.5 mM KCl, 1.25 mM NaH$_2$PO$_4$, 10 mM D-glucose, 2 mM MgCl$_2$, 2 mM CaCl$_2$ in tissue culture-grade ddH$_2$O.

2. Dissection buffer: HBSS (Gibco, Cat# 14175-095) supplemented with 100 U/mL penicillin and 100 μg/mL streptomycin.

3. Digestion buffer: 2.5 U/ml dispase (Dispase II: Roche Applied Science, Cat# 049420780010) and 200 U/ml collagenase (Worthington Biochemical, CLS2, Cat#LS004176) in HBSS (Gibco, Cat#14025-092). Stored at −80 °C in 1 ml aliquots for up to 6 months.

4. Neuron feeding medium: Neurobasal A media (Gibco, Cat# 10888-022) supplemented with 2% B27 (Gibco, Cat# 17504-044), 1% GlutaMax (Gibco, Cat#35050-061), 1% Anti-Anti (Gibco, Cat#15240-062), 2% heat-inactivated fetal bovine serum (Hyclone, Cat#SH30071.03). Media filtered using a corning filter system (Cat# 430320).

5. Poly-L-ornithine solution: Sigma, P4957-50 ml.

6. Laminin: Roche, Cat#11243217001.

7. 2.5% Avertin solution: 2,2,2-Tribromoethanol (Sigma, Cat#75-80-9) and 2-methyl-2-butanol (Sigma, Cat#75-85-4) dissolved in tissue culture-grade ddH$_2$O.

2.1.2 Dissection Tools and Other Materials

1. Forceps: 1 pair, Dumont #5/45, No.11251-35, Fine Science Tools.

2. Spring scissor: Cohan-Vannas, No.15000-01, Fine Science Tools.

3. Pre-cleaned glass coverslips: 25 mm Deckgläser coverslips (Carolina Biosciences Cat# 633037).

4. Cell strainer: Fisherbrand, 70 μm nylon mesh, Cat#22363548.

2.2 Primary Mouse DRG Neuron Transfection

1. Basic nucleofector kit for primary mammalian neurons (SCN): Lonza, Cat#VSPI-1003.

2. Amaxa Nucleofector II (Lonza).

3. GFP-LC3 constructs and any other constructs or siRNA if co-expression is needed.

2.3 Live-Cell Imaging of Autophagosomes in Primary DRG Neurons

1. Imaging chamber: ThermoFisher, Cat#A7816.

2. Neuronal imaging media: hibernate a low fluorescence medium (BrainBits) supplemented with 2% B27 (Gibco, Cat# 17504-044) and 1% GlutaMax (Gibco, Cat#35050-061).

3. Confocal microscope (LSM 880; Carl Zeiss) equipped with an environmental chamber.

2.4 Analysis of Autophagosome Dynamics

1. ImageJ.

2. GraphPad Prism 7.

3 Methods

3.1 Primary Adult Mouse DRG Neuron Culture, Nucleofection, and Maintenance

1. The day before neuron culture, coat methanol/EtOH-washed glass coverslips with poly-L-ornithine solution diluted with PBS (1:4) and incubate at 37 °C in tissue culture incubator overnight. Prior to dissection the next day, remove the poly-L-ornithine solution and rinse three times with sterile water (5 min each). Remove water, coat coverslips with Laminin (1:50 in PBS), and incubate in tissue culture incubator during the dissection (minimum of 1 h). Remove the Laminin solution and rinse the coverslip once with sterile water before plating the neurons.

2. Before dissection, oxygenate ACSF (60 ml for each mouse) by placing the container in an ice box and incubating at 4 °C for at least 20 min.

3. Anesthetize the mouse by intraperitoneal injection of 2.5% avertin, usually 25 g for 0.5 ml. After the mouse is fully anesthetized, perfuse with oxygenated ACSF. A pale-colored liver indicates a successful perfusion.

4. Dissection (*see* **Note 1**):

   (a) Using pins on a foam board, stabilize the mouse's position with its ventral side facing downward. Decapitate and then grab hold of the mouse's back skin with one tweezer, making an incision through the skin from tail to neck with a small scissor in order to expose the connective tissue underneath. Pull the skin away and position on one side.

   (b) Cut off the shoulder blade and remove adipose tissue, connective tissue, and muscle on the spinal column from tail to neck with a small scissor in order to expose the bone structure of the entire spinal column.

   (c) Use sharp-pointed scissors to make a cross-sectional incision of the spinal column at the tail side. Place the blade of the scissors through the incision into the spinal column, cutting through each side of the dorsal spinal vertebrae from tail to neck exposing the spinal cord. While cutting, it is important to maintain a 30° angle above the middle plane and above the row of DRGs.

   (d) To remove the ventral half of the spinal column, take hold of its caudal side and cut the muscles connected to it. Gently wash the removed spinal column in a 60-mm petri dish containing ice-cold dissection buffer and place it on an ethanol-cleaned ice pack under the dissection microscope.

   (e) DRGs are connected by two roots—the anterior root and posterior root, both stemming from the spinal cord. Under the dissection microscope, use a pair of fine forceps to remove the DRGs one at a time from tail to neck starting on one side. First, push the spinal cord to the other side to expose the axon roots connecting the spinal cord and DRGs, which localize in the intervertebral foramina ("the pocket"). Use one forceps to gently pull and hold the posterior root, allowing the DRG to move outward. Use the other forceps to pinch the anterior root at the other side of the ganglion, thereby pulling the ganglion out from the pocket and off of the spinal cord (*see* **Note 2**). Transfer the DRG to a 35-mm petri dish containing cold dissection buffer (maintained on an ice pack).

Continue until all the DRGs from both sides have been collected. About 30–40 DRGs can be collected from each mouse if the operator is skilled and careful when exposing the cord.

(f) Use fine forceps and spring scissors to trim off the excess roots and other nonneuronal tissue (e.g., blood) attached to each ganglion, getting as close to the ganglion as possible. Transfer the trimmed DRGs to a 1.5-ml EP tube containing 1 ml ice-cold digestion buffer.

5. Dissociation:

(a) Incubate the ganglia in digestion buffer (dispase/collagenase) in the EP tube for 28–30 min at 37 °C in a tissue culture incubator. Place the tube horizontally so that the ganglia can be distributed evenly. Gently shake the tube every 10 min to prevent the ganglia from sticking to each other. Afterwards, place the EP tube on a rotator at room temperature for 35 min with gentle shaking.

(b) Centrifuge the EP tube at 950 rpm for 4 min at room temperature.

(c) Carefully remove the supernatant using a 1-ml pipette tip.

(d) Add 1 ml of pre-warmed neuron feeding medium and mechanically dissociate the ganglia by gently triturating no more than six times with a 1-ml pipette tip.

(e) Let any large blocks of DRGs to settle to the bottom of the EP tube. In the meantime, prepare a 70- μm cell strainer by placing it on top of a 50-ml sterile corning tube.

(f) Carefully collect the supernatant while leaving any settled blocks of DRGs undisturbed. Place the supernatant on top of the 70- μm cell strainer allowing it to flow into the 50-ml tube. Add another 1 ml of pre-warmed neuron feeding medium to the EP tube containing the remaining blocks of DRGs. Gently triturate six times with a 1-ml pipette tip. Transfer this entire mixture to the same 70- μm cell strainer on top of the 50-ml tube.

(g) Add 3 ml of pre-warmed neuron feeding medium to the strainer to wash any remaining neurons trapped by the strainer into the 50-ml tube. Discard the strainer containing any remaining large blocks of undissociated tissue.

(h) Remove the strainer from the tube and transfer 5 ml of medium containing DRG neurons into a 15-ml tube. Count cells with a hemocytometer. It is expected that 1–1.5×10^6 cells can be harvested from 30 to 40 DRGs isolated from one mouse. Importantly, in order to proceed to nucleofection, there must be at least 2×10^4 cells for each transfection.

6. Nucleofection:

    (a) Centrifuge the 15-ml tube at $100 \times g$ for 10 min.

    (b) Carefully remove the medium using gentle vacuum aspiration. Use a 200 µl tip to remove any remaining medium close to the cell pellet.

    (c) Resuspend the cell pellet with 20 µl nucleofection solution using a 200- µl pipette tip to break up the cell pellet. Add 0.3 µg GFP-LC3 DNA construct for each transfection with or without other DNA/siRNA constructs according to the experimental design.

    (d) Gently mix and transfer the solution to the transfection cuvette. Once the cuvette is inside the cuvette holder of the nucleofector, pulse once using the SCN Basic Neuro Program 6 in the cell type list.

    (e) Immediately after transfection, allow neurons to recover by adding 200 µl of pre-warmed medium to the cuvette and placing the cuvette in a 37 °C tissue culture incubator for 5–10 min.

7. Gently mix the cells in the cuvette and count the cell number using a hemocytometer. Using neuron feeding medium to dilute the cells, plate the DRG neurons at an optimized density. For examining autophagosome dynamics with confocal microscopy, 7000–10,000 cells/25 mm^2 coverslip maintained in the well of a 6-well plate with 2 ml medium is recommended.

8. Note that a small amount of myelin and fibrous debris may be present in the medium. Completely change the medium 3–4 h after plating the cells. To maintain the cells, change half of the medium every other day (*see* **Note 3**).

3.2 Time-Lapse Live-Cell Imaging of Autophagosomes in Primary DRG Neuron Axons

1. To capture autophagosome dynamics, DRG neurons can be imaged as early as DIV2. Prior to imaging, DRG neurons should be starved by exchanging to FBS-free media for 2–3 h (*see* **Note 4**).

2. Equilibrate environmental chamber, as well as the microscope stage insert, to 37 °C at least 30 min before imaging.

3. Place coverslip in imaging chamber and add 1 ml pre-warmed neuronal imaging media. Place imaging chamber in the 37 °C insert. Make sure to use an insert with a cover to prevent media from evaporating quickly.

4. Cultured DRG neurons have multiple neurites, all of which are axons (Fig. 1). The axons are well extended, smooth, and branched at the terminal. When selecting cells for imaging, avoid neurons that are not well developed or appear to have bulbous structures along the axon (*see* **Note 5**).

5. When autophagy has not been initiated, LC3 remains cytosolic, and thus, the GFP signal will be diffused throughout the axons. After a 3-h neuronal starvation, autophagy is induced above basal levels, and LC3 gradually transforms into the lipidated form and is recruited to the double membrane of autophagosomes, resulting in GFP signal to appear as discrete puncta within the neurons (*see* **Note 6**). Autophagosomes are considered as GFP-LC3-positive punctate structures.

6. Autophagosome biogenesis is usually enriched in the distal axon (~100 μm to the terminal), especially in the growth cone. A practical strategy for choosing an axon to image for autophagosome dynamics is to look for GFP puncta at the terminal. Our observations suggest that an axon will less likely contain autophagosomes in the proximal or middle regions if its terminal has no detectable autophagosomes.

7. Autophagosomes exhibit bidirectional transport in the very proximal and very distal regions of axons. To record long-distance motility of autophagosomes, choose the middle region of the axon (~100 μm from the soma and growth cone, respectively) and note the direction of the cell body.

8. Use a 40× or 63× objective to capture a time series according to the following guidelines:
    (a) At a minimum, acquire image frames every 2 s for at least 3 min. If dual-color time-lapse imaging is required, e.g., GFP-LC3 co-expression with the late-endosome marker RFP-Rab7, choose the "Line Scan" function to avoid time lag between the two channels.

    (b) If the GFP signal is too dim, first try adjusting the average number and detection gain, instead of laser power and pixel dwell time, in order to minimize bleaching the GFP signal over time.

    (c) Use the "Definite Focus" function to maintain focal plane during image acquisition.

9. For achieving consistent data, do not image an individual coverslip for more than 3 h, as declining neuronal health may cause variability beyond this point (*see* **Note 7**).

3.3 Analysis of Axonal Autophagosome Dynamics

1. Open microscope image files (usually present as "LSM") in ImageJ (NIH). GFP-LC3 often has a high background, thus you may use the "subtract background" macro to reduce background signals prior to quantification.

2. Though GFP-LC3 appears as punctate structures along an axon, sometimes there will be non-autophagosome green dots (e.g., nonspecific green dots in the background or a small protrusion of the axon enriched with GFP). To exclude these artifacts, play the movie to determine whether individual

green puncta represent authentic autophagosomes. The trajectory of an autophagosome will remain within the axon, while nonspecific dots often reside outside the axonal boundary.

3. Axonal autophagosome density:

    (a) To quantify the density of autophagosomes along an axon, first count the number of GFP-LC3 puncta along the axon. To determine if a larger structure is a single autophagosome or multiple individual puncta, play the movie to identify multiple autophagosomes that move independently, collide, or separate from one another.

    (b) Trace the axon with the "segmented line" feature to measure the length (μm).

    (c) Normalize the number of autophagosomes by the axon length to obtain "number of autophagosomes per 100 μm."

4. Axonal autophagosome motility:

    (a) Motility analysis is used to quantify the percentage of autophagosomes moving in the anterograde direction, those moving in the retrograde direction, and those remaining stationary. For consistency and clarity, we always position the cell body on the left side of the image series so that the direction of autophagosome motility will be very clear.

    (b) If the signal diminishes as the image series progresses, run the "bleach correction" plug-in.

    (c) We use a macro written in-house to generate our kymographs [27]. The code is saved in "Startup Macros" as follows: (To editor: The line spacing of the macro below is too wide, which should be single line spacing just like introduction part)

```
macro "Make Kymograph... [k]" {
run("32-bit");
setTool(7);
run("Straighten ");
title = "Trace";
msg = "just trace, press straighten, then stack and OK";
waitForUser(title, msg);
//run("Reslice [/]...", "input=4 output=1.000 start=Top");
selectImage(1)
getPixelSize(unit, pw, ph, pd);
if (nSlices<=30) {
myParameter="input=" + 1/pw + " output=1.000 start=Top";
} else {
myParameter="input=" + 1/pw + " output=1.000 slice=21";
}
```

```
selectImage(4);
run("Reslice [/]...", myParameter);
//print("updated myReSlice: "+myParameter);
selectImage(3);
close();
selectImage(2);
close();
selectImage(2);
close();
selectWindow("Reslice of Straightened");
getLut(reds, greens, blues);
if (reds[10]>greens[10]) {
run("RGB Color");
// run("Grouped ZProjector", "group=100 projection=[Max In-
tensity]");
//if (nSlices>=30) {
// Title=getTitle();
//index2=indexOf(Title, ".lsm Channel");
//myTitle=substring(Title, 0, index2);
//}
mySliceNum=nSlices();
myParameter="group=" + toString(mySliceNum) + " projection=
[Max Intensity]";
run("Grouped ZProjector", myParameter);
//print("updated mySliceNum: "+mySliceNum);
//print("updated myParameter: "+myParameter);
rename("Red channel kymograph");
run("Size...", "width="+getWidth()+" height=2000 interpola-
tion=Bilinear");
selectWindow("Reslice of Straightened");
run("Close");
selectWindow("Red channel kymograph");
Dir=getDirectory("Choose a Directory");
saveAs("Jpeg", Dir+"Red kymograph");
}
else if (reds[10]<greens[10]) {
run("RGB Color");
// run("Grouped ZProjector", "group=100 projection=[Max In-
tensity]");
//if (nSlices>=30) {
//Title=getTitle();
//index2=indexOf(Title, ".lsm Channel");
//myTitle=substring(Title, 0, index2);
//}
mySliceNum=nSlices();
myParameter="group=" + toString(mySliceNum) + " projection=
[Max Intensity]";
run("Grouped ZProjector", myParameter);
```

```
//print("updated mySliceNum: "+mySliceNum);
//print("updated myParameter: "+myParameter);
rename("Green channel kymograph");
run("Size...", "width="+getWidth()+" height=2000 interpola-
tion=Bilinear");
selectWindow("Reslice of Straightened");
run("Close");
selectWindow("Green channel kymograph");
Dir=getDirectory("Choose a Directory");
saveAs("Jpeg", Dir+"Green kymograph");
}
else {
run("RGB Color");
// run("Grouped ZProjector", "group=100 projection=[Max In-
tensity]");
//if (nSlices>=30) {
//Title=getTitle();
//index2=indexOf(Title, ".lsm Channel");
//myTitle=substring(Title, 0, index2);
//}
mySliceNum=nSlices();
myParameter="group=" + toString(mySliceNum) + " projection=
[Max Intensity]";
run("Grouped ZProjector", myParameter);
//print("updated mySliceNum: "+mySliceNum);
//print("updated myParameter: "+myParameter);
rename("Kymograph");
run("Size...", "width="+getWidth()+" height=2000 interpola-
tion=Bilinear");
selectWindow("Reslice of Straightened");
run("Close");
selectWindow("Kymograph");
Dir=getDirectory("Choose a Directory");
saveAs("saveAs("Jpeg", Dir+Name+" Dual channel kymograph");
}
```

(d) If an axon fails to remain in the same position during the imaging and shifted a small distance, play the whole movie to estimate the distance of the shift. Set the number of pixels in the "width of Filament/Wide Line" large enough to cover the shift after tracing the axon.

(e) Adjust the brightness and contrast to make the generated kymograph more clear for presentation.

(f) Using the kymograph, count the number of autophagosomes moving in each direction. As the cell body is always positioned on the left of our image series, kymograph analysis can be easily summarized: slanted lines or curves to the

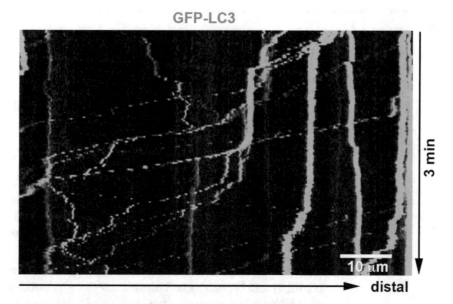

Fig. 3 Axonal autophagosomes present a dominant retrograde motility. Kymograph based on time-lapse live-cell images showing the majority of GFP-LC3-labeled autophagosomes moving in retrograde direction along the axon. Each line represents a single autophagosome. Vertical lines represent stationary organelles; diagonal lines to the right represent anterograde movement; to the left indicate retrograde movement. An autophagosome is considered stationary if it remains immotile with net displacement ≤10 μm. Adult mouse DRG neurons were transfected with GFP-LC3 at DIV0 and imaged at DIV3 after incubation with serum-free medium for 3 h. Time-lapse images were acquired from the middle segment of axons. Scale bar: 10 μm

right (negative slope) represent anterograde movement; those to the left (positive slope) indicate retrograde movement. Vertical lines represent stationary organelles (Fig. 3). An autophagosome is considered stationary if it remains immotile (net displacement ≤10 μm) during entire recording time.

(g) Quantify the percentage of autophagosomes in each direction based on the total number of autophagosomes imaged.

5. For both autophagosome density and motility analysis, each data point is obtained from a single axon that can be easily traced back to a single neuron. Measure at least 30–35 individual neurons from three independent experiments for each assay.

6. Perform statistical analysis using GraphPad Prism 7 (*see* **Notes 8** and **9**).

4   Notes

1. For each adult mouse, dissection should be performed in less than 1 h. Longer duration not only leads to difficulties pulling the ganglion off the spinal cord since the tissue gets soft over time, but also impairs the overall neuron quality and survival rate.

2. Since neuronal cell bodies reside within the ganglion, extreme care should be taken not to pinch the ganglion with your forceps when trying to pull it from the pocket.

3. DRG axons should extend from the cell body beginning at DIV1. If there are no axons extending from the DRG at this early stage, this particular neuron is probably unhealthy and unlikely to grow axons later.

4. Though starving cells by FBS removal from the medium is a routinely used protocol to induce autophagy in DRG neurons, alternative methods (e.g., treating cells with rapamycin, Torin1 or removing amino acids from the medium) can also be used if necessary.

5. DRG neurons with medium-sized cell bodies are more likely to exhibit dynamic autophagy events.

6. DRG neurons with too bright or too dim GFP signal should not be selected for imaging, as these cells are frequently characterized by poorly developed neurites or aberrant morphological branches, making it very difficult to observe autophagic events even after 3-h starvation.

7. Always perform control experiments during the autophagosome dynamics assay. The autophagosome transport pattern in a healthy neuron should present a dominant retrograde motility.

8. The autophagosome imaging and analysis protocol can be applied to study dendritic and axonal autophagosome dynamics in other neuronal cell types (e.g., cortical neurons, hippocampal neurons, motor neurons) with GFP-LC3 expression.

9. For examining axonal autolysosomes and assessing autophagic clearance in degradative lysosomes, special precautions should be considered. Although lysosomal-associated membrane protein 1 (LAMP1) and LAMP2 target to lysosomes and have been routinely used as lysosome markers, they are not static components of degradative lysosomal membranes. Instead, LAMP1/2 are in dynamic equilibrium between endosomes, lysosomes, amphisomes, and autolysosomes. Our recent study indicated that LAMP1/2 are neither specific markers to assess degradative lysosome distribution and trafficking in neurons nor sensitive indicators to reveal the pathological response of

the autophagy-lysosome system in some neurodegenerative diseases [28]. We suggest that labeling a set of lysosomal hydrolases combined with various autophagic markers would be more accurate than simply relying on LAMP1/2 staining to assess neuronal lysosome and autophagic degradation under physiological and pathological conditions.

Acknowledgments

This work was supported by the Intramural Research Program of NINDS, NIH ZIA NS002946, and ZIA NS003029 (Z-H. S.).

References

1. Nixon RA (2013) The role of autophagy in neurodegenerative disease. Nat Med 19:983–997

2. Maday S (2016) Mechanisms of neuronal homeostasis: autophagy in the axon. Brain Res 1649:143–150

3. Ariosa AR, Klionsky DJ (2016) Autophagy core machinery: overcoming spatial barriers in neurons. J Mol Med (Berl) 94:1217–1227

4. Levine B, Klionsky DJ (2004) Development by self-digestion: molecular mechanisms and biological functions of autophagy. Dev Cell 6:463–477

5. Shen HM, Mizushima N (2014) At the end of the autophagic road: an emerging understanding of lysosomal functions in autophagy. Trends Biochem Sci 39:61–71

6. Tooze SA, Abada A, Elazar Z (2014) Endocytosis and autophagy: exploitation or cooperation? Cold Spring Harb Perspect Biol 6: a018358

7. Hara T et al (2006) Suppression of basal autophagy in neural cells causes neurodegenerative disease in mice. Nature 441:885–889

8. Komatsu M et al (2006) Loss of autophagy in the central nervous system causes neurodegeneration in mice. Nature 441:880–884

9. Komatsu M et al (2007) Essential role for autophagy protein Atg7 in the maintenance of axonal homeostasis and the prevention of axonal degeneration. Proc Natl Acad Sci U S A 104:14489–14494

10. Nishiyama J et al (2007) Aberrant membranes and double-membrane structures accumulate in the axons of Atg5-null Purkinje cells before neuronal death. Autophagy 3:591–596

11. Zhou Z et al (2015) Autophagy supports survival and phototransduction protein levels in rod photoreceptors. Cell Death Differ 22:488–498

12. Gowrishankar S et al (2015) Massive accumulation of luminal protease-deficient axonal lysosomes at Alzheimer's disease amyloid plaques. Proc Natl Acad Sci U S A 112:E3699–E3708

13. Karabiyik C, Lee MJ, Rubinsztein DC (2017) Autophagy impairment in Parkinson's disease. Essays Biochem 61:711–720

14. Lee S, Sato Y, Nixon RA (2011) Lysosomal proteolysis inhibition selectively disrupts axonal transport of degradative organelles and causes an Alzheimer's-like axonal dystrophy. J Neurosci Off J Soc Neurosci 31:7817–7830

15. Schneider JL, Cuervo AM (2014) Autophagy and human disease: emerging themes. Curr Opin Genet Dev 26:16–23

16. Xie Y et al (2015) Progressive endolysosomal deficits impair autophagic clearance beginning at early asymptomatic stages in fALS mice. Autophagy 11:1934–1936

17. Zheng S et al (2010) Deletion of the huntingtin polyglutamine stretch enhances neuronal autophagy and longevity in mice. PLoS Genet 6:e1000838

18. Bolam JP, Pissadaki EK (2012) Living on the edge with too many mouths to feed: why dopamine neurons die. Mov Disord 27:1478–1483

19. Matsuda W et al (2009) Single nigrostriatal dopaminergic neurons form widely spread and highly dense axonal arborizations in the neostriatum. J Neurosci 29:444–453

20. Cai Q et al (2010) Snapin-regulated late endosomal transport is critical for efficient autophagy-lysosomal function in neurons. Neuron 68:73–86

21. Sheng ZH, Cai Q (2012) Mitochondrial transport in neurons: impact on synaptic

homeostasis and neurodegeneration. Nat Rev Neurosci 13:77–93

22. Maday S, Wallace KE, Holzbaur EL (2012) Autophagosomes initiate distally and mature during transport toward the cell soma in primary neurons. J Cell Biol 196:407–417

23. Cheng XT et al (2015a) Axonal autophagosomes recruit dynein for retrograde transport through fusion with late endosomes. J Cell Biol 209:377–386

24. Ferguson SM (2018) Neuronal lysosomes. Neurosci Lett 697:1–9

25. Owen DE, Egerton J (2012) Culture of dissociated sensory neurons from dorsal root ganglia of postnatal and adult rats. Methods Mol Biol 846:179–187

26. Cheng XT et al (2015b) Axonal autophagosomes use the ride-on service for retrograde transport toward the soma. Autophagy 11:1434–1436

27. Di Giovanni J, Sheng ZH (2015) Regulation of synaptic activity by snapin-mediated endolysosomal transport and sorting. EMBO J 34:2059–2077

28. Cheng XT et al (2018) Characterization of LAMP1-labeled nondegradative lysosomal and endocytic compartments in neurons. J Cell Biol 217:3127–3139

Chapter 9

Imaging and Quantifying Neuronal Autophagy to Determine the Autophagy Contribution to Neuronal and Dendritic Morphogenesis

Wan Yun Ho and Shuo-Chien Ling

Abstract

Autophagy is one of the critical processes for the cells to maintain proper homeostasis by eliminating damaged organelles and unwanted macromolecules as well as recycling the building blocks. Autophagy inactivation has been shown to cause neurodegeneration, and autophagy dysfunction is linked with neurodegenerative diseases. Furthermore, accumulating evidence indicates that autophagy is involved in synaptic biology and plasticity, and the dysfunctions of "synaptic autophagy" may also contribute to neuronal dysfunction as part of the neurodegeneration process. In this context, the role of autophagy in neuronal morphogenesis, synapse formation, and maintenance remains to be clarified. Here, we describe methods to image and quantify neuronal autophagy and their contributions to neuronal and dendritic morphogenesis using primary rodent hippocampal and cortical neuron cultures. We use C9ORF72, the most common genetic cause for amyotrophic lateral sclerosis (ALS) and frontotemporal dementia (FTD), as an example to illustrate how autophagy is required for neuronal and dendritic morphogenesis.

Keywords Amyotrophic lateral sclerosis (ALS), Frontotemporal dementia (FTD), C9ORF72, Autophagy, LC3

1 Introduction

Autophagy is a highly conserved and regulated bulk degradation pathway that degrades unused or dysfunctional cytoplasmic material and organelles, preparing for the turnover and recycling of cellular constituents. Therefore, the key physiological functions of autophagy are to (1) provide nutrients essential for survival in response to nutrient deprivation and (2) eliminate unnecessary components, including protein aggregates. These functions are important for not only cellular remodeling during differentiation and development, but also maintaining cellular and organismal homeostasis [1, 2]. There are three different types of autophagy, depending on the route by which cytoplasmic material can be transported to the lysosomal vesicles: macroautophagy,

Ben Loos and Esther Wong (eds.), *Imaging and Quantifying Neuronal Autophagy*, Neuromethods, vol. 171,
https://doi.org/10.1007/978-1-0716-1589-8_9, © Springer Science+Business Media, LLC, part of Springer Nature 2022

microautophagy, and chaperone-mediated autophagy. In macroautophagy, the formation of a double-membraneous isolation membrane targets a portion of the cytoplasm for degradation. The endosome of the targeted cytoplasm for degradation is called autophagosome. The autophagosome then fuses with a lysosome to form the autolysosome. Lysosomal hydrolases in the autolysosome would then degrade the cytoplasm as well as the inner limiting membrane. On the other hand, microautophagy involves the invagination of the lysosomal membrane to engulf a region of the cytosol. Lastly, in chaperone-mediated autophagy, selected cytosolic proteins bound to chaperones are targeted to the lysosome for degradation. Regardless of which mode of autophagy is undertaken by the cell, the degraded products are subsequently recycled back to the cytoplasm for biosynthesis or energy production [1, 3].

Among three different types of autophagy, macroautophagy (referred to hereafter as autophagy) can be grouped into distinct phases with specific complexes executing defined functions. Specifically, induction of autophagy starts with the activation of the autophagy initiation complex that is composed of ULK1, FIP200/RB1CC1, ATG13, and ATG101. Next, ULK1 phosphorylates the class III phosphoinositide 3-kinase (PI3K-III) complex containing BECN1 (Beclin 1), ATG14L1, Ambra1, VPS34, and VPS15, which nucleates phagophore formation. Subsequently, two ubiquitin-like conjugation systems involved in ATG12-ATG5-ATG16L1 and LC3/ATG8-lipidation mediate the elongation and closure of the phagophore membrane to form the double-membrane-containing autophagosome, which fuses with a lysosome in the final step of autophagic degradation [4, 5].

In many age-related neurodegenerative diseases, there is notable accumulation of ubiquitin-positive protein aggregates in affected regions of the brain [6]. Failing protein homeostasis, including autophagy dysfunctions, not only associates with aging process, but also leads to pathological protein inclusions that are hallmarks of many neurodegenerative diseases, including Alzheimer's disease (AD), Parkinson's disease (PD), amyotrophic lateral sclerosis (ALS), and frontotemporal dementia (FTD) [7, 8]. Although it remains to be debated whether these misfolded and aggregated proteins per se disrupt neuronal function and subsequently lead to neurodegeneration, the inability to properly maintain protein homeostasis clearly contributes to neurodegeneration. Studies conducted by Komatsu et al. and Hara et al. show that loss of autophagic function in the central nervous system of mice caused the accumulation of ubiquitin-positive inclusions in the brain tissues through genetic inactivation of key autophagy genes (Atg7 and Atg5, respectively) and consequently neurodegeneration [9, 10]. Furthermore, the enhancement of autophagy in animal models of neurodegeneration improves clearance of protein aggregates and reduces symptoms of neurodegeneration

[11, 12]. Collectively, the evidence demonstrates that proper autophagic function is essential for the health of the nervous system, and autophagy dysfunction contributes to neurodegeneration.

In contrast, the role of autophagy in neuronal morphogenesis, synaptogenesis, synapse maintenance, and synaptic plasticity is just being established. For example, two independent autophagy inhibitors (SBI-0206965 and Spautin-1), which target ULK1 and Beclin 1, respectively, reduce dendritic arborization in primary hippocampal neurons [13]. Similarly, pharmacological inhibition of autophagy using 3-methyladenine (3-MA) reduces neurite outgrowth in the adult dorsal root ganglia (DRG) neurons [14]. However, genetic reduction of Atg7 by RNAi enhances the axon extension in rat primary cortical neurons [15]. Work from the Holzbaur laboratory showed that autophagosome biogenesis occurs in axons and rarely in soma and dendrites in neuronal culture [16]. The presynaptic autophagy appears to be important for synapse formation at least in *C. elegans* [17] and *Drosophila* [18]. Furthermore, components of the autophagy machinery interact with presynaptic proteins, such as endophilin A and Bassoon, to induce or inhibit autophagy [19, 20]. Interestingly, neuronal activities appear to regulate autophagy and may impact synaptic plasticity [21]. Furthermore, mTOR-dependent autophagy is required for synapse pruning [22]. Autophagy inhibition reduces spine density in the hippocampal neurons [13]. Taken together, although autophagy participates in neuronal morphogenesis and synaptic biology, the exact mechanisms remain to be clarified.

In this chapter, we describe methods to image and quantify neuronal autophagy and their contributions to neuronal and dendritic morphogenesis using primary rodent hippocampal and cortical neuron cultures. We use C9ORF72, the most common genetic cause for ALS and FTD [23, 24], as an example to illustrate how autophagy is required for neuronal and dendritic morphogenesis.

2 Materials

2.1 Primary Hippocampal and Cortical Neuron Culture

Besides the advance using human iPSCs (induced pluripotent stem cells) and differentiating them into different neuronal lineages, such as cholinergic, glutamatergic, dopaminergic, or motor neurons, dissociated primary hippocampal neuron cultures and mixed cortical neuron cultures are well-established and well-accepted culture system [25, 26]. In particular, the primary hippocampal neuron culture has been used extensively for investigating neuronal morphogenesis and synaptogenesis [27–30].

Wash coverslips once with 10 ml of 70% ethanol in a 10 cm dish, followed by three quick rinses with 10 ml of autoclaved ddH$_2$O.-Place a washed coverslip into each of the well of a 24-well plate, dry

2.1.1 Clean and Treat Glass Coverslips with Poly-L-Lysine (for Hippocampal Neuron Culture)

coverslips in wells thoroughly. Dilute poly-L-lysine (Sigma-Aldrich, P4707) at 30 µg/ml in 0.1 M borate buffer, pH 8.5. Add 500 µl poly-L-lysine solution into each well. Incubate at 37 °C overnight. Remove the poly-L-lysine and wash each well three times with autoclaved ddH$_2$O. Let the coverslips in wells dry completely in the laminar flow hood for at least 4 h prior to use.

2.1.2 Treat Glass 6-Well Cell Culture Plated with Poly-L-Lysine (for Cortical Neuron Culture)

Dilute poly-L-lysine at 30 µg/ml in 0.1 M borate buffer, pH 8.5. Add 2 ml poly-L-lysine solution into each well. Incubate at 37 °C overnight. Remove the poly-L-lysine and wash each well three times with autoclaved ddH$_2$O. Let the coverslips in wells dry completely in the laminar flow hood for at least 4 h prior to use.

2.2 Constructs for Visualization of Neuronal Morphology and Autophagy

The neurons were transfected with plasmids encoding EGFP or GFP^{NLS}-P2A-mCherry^{CAAX} to visualize dendritic spines using calcium phosphate precipitation or Lipofectamine 3000 (Thermo Fisher Scientific, L3000015) at DIV13 [13].

2.3 Autophagy Induction and Inhibition with the Use of Small Molecules

Many autophagy inducers and inhibitors are readily available from commercial source. Readers are encouraged to refer to the excellent guideline [31].

2.4 Genetic Manipulation

With the public availability of genetically modified animals and advancement of knockdown and knockout approaches, such as siRNA and CRISPR-Cas9, it is straightforward to examine neuronal autophagy by inactivating autophagy genes. For example, c9orf72 knockout mice were generated as described previously [13, 32] and maintained in a C57BL/6 background. For pooling animals for neuronal culture, non-transgenic and c9orf72 homozygous knockout pups are generated by mating a pair of non-transgenic C57BL/6 and homozygous c9orf72 knockout mice, respectively.

3 Methods

3.1 Hippocampal and Cortical Neuron Culture from Wild-Type and Genetically Modified Mice

Dissociated primary hippocampal neuron culture and mixed cortical neuron cultures have been used extensively for investigating neuronal morphogenesis and synaptogenesis, and synapse maintenance [25–30]. It has been shown that the dissociated primary hippocampal neuron cultures contain 80–90% of glutamatergic excitatory neurons and 10–20% of GABAergic inhibitory neurons [25, 26]. Thus, the observed changes are likely due to the majority of the glutamatergic pyramidal and granule cells. In addition, by day in vitro (DIV) 13, all major branches are dendrites with visible dendritic spines with a thin and long axon [28] (*see* **Note 1**).

With the availability of extensive genetic engineered mouse models and the genetic (i.e., siRNA and CRISPR-Cas9, etc.) and pharmacological tools, the system is well suited to study how autophagy may affect the neuronal morphogenesis and synaptic biology. For example, by mating hemizygous C9orf72 knockout mice (C9orf72$^{+/-}$) [13], the primary hippocampal neurons from the three genotypes representing wild-type (C9orf72$^{+/+}$), C9orf72 knockout hemizygote (C9orf72$^{+/-}$), and C9orf72 knockout homozygote (C9orf72$^{-/-}$) can be obtained (Fig. 1a). Using GFP as a cell filler, the neuronal morphology and dendritic spines can be visualized (Fig.1b, c) and subsequently quantified (Fig. 1d, e). Consistently with our previous finding [13], C9orf72 is required for neuronal morphogenesis and maintaining spine density in a dose-dependent manner (Fig. 1).

3.1.1 Dissection of Hippocampi from P0 Mouse Pups

Sterilize all dissecting instruments by washing them with 70% ethanol. Euthanize the pup by decapitation using a scissors to separate the head from body. Dissect the skin and skull on the top of the head and hold down the tissues on either side using forceps. By using a spatula, gently lift the brain from the base and transfer it to a 60 mm dish containing ice-cold Hank's Balanced Salt Solution (HBSS) washing buffer, which comprises of 1% penicillin-streptomycin (Thermo Fisher Scientific, 15070063) and 10 mg/ml of dextrose (Sigma-Aldrich, G8270) in 1× HBSS (Life technologies, 14185-052). Separate the two hemispheres by making a sagittal cut along the midline. Orientate the brain such that the outer hemisphere touches the bottom of the dish. Under the dissecting microscope, remove the midbrain tissues using fine forceps carefully, leaving an intact hemisphere containing the cortex and hippocampus. Turn the brain over gently so that the hippocampus touches the bottom of the dish. Hold the hemisphere in place by pinning down the olfactory bulb with a pair of forceps. With another pair of forceps, gently pinch and peel off the meninges and remove them completely. Reorientate the tissue such that the hippocampus faces upward. Using a fine pair of scissors, outline and remove the hippocampus carefully. Transfer the tissues to a 15 ml tube containing 5 ml of ice-cold HBSS washing buffer. For pooled hippocampal culture, place hippocampi from three pups into one tube. For cortical culture, place cortices from one pup into one tube.

3.1.2 Cell Dissociation and Plating of Neuronal Cells

Let the tissues settle down to the bottom and aspirate the wash buffer. Add 10 ml of fresh ice-cold HBSS washing buffer, wait for tissues to settle, aspirate and repeat twice. Add 5 ml of pre-warmed 0.125% trypsin (Biowest, L0931-100) in 1× HBSS, swirl tube to gently mix. Incubate at 37 °C for 15 min. Stop trypsin activity by

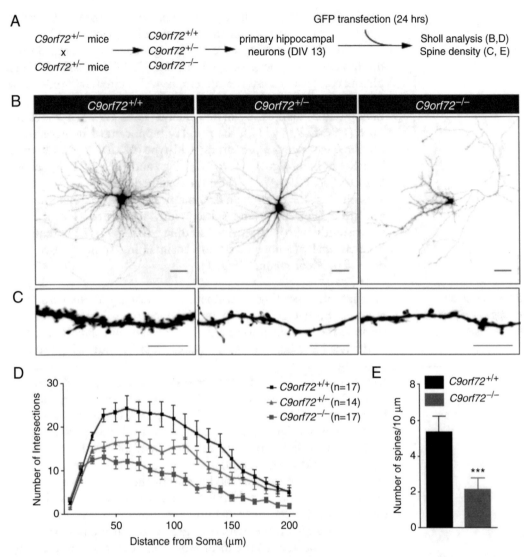

Fig. 1 C9ORF72 dose-dependent reduction on dendritic arborization in primary hippocampal neurons. (**a**) Schematic of experimental design. Hemizygous *c9orf72* knockout mice (C9orf72$^{+/-}$) were crossed with C9orf72$^{+/-}$ to generate wild-type (C9orf72$^{+/+}$), C9orf72 knockout hemizygote (C9orf72$^{+/-}$), and C9orf72 knockout homozygote (C9orf72$^{-/-}$). Primary hippocampal neurons were cultured from each genotype and transfected with GFP at DIV (days in vitro) 13 to visualize the neuronal and spine morphology. (**b**) Images of GFP-transfected hippocampal neurons from C9orf72$^{+/+}$, C9orf72$^{+/-}$, and *c9orf72*$^{-/-}$ mice. The images are presented in gray scale and inverted color. Scale bar: 50 μm. (**c**) High-magnification renderings of dendritic segments. The dendritic images were acquired using a 0.3- μm step width in *z* and then stacked as a maximum projection. Scale bar: 10 μm. (**d**) Sholl analysis of dendritic arborization. (**e**) Reduced spine density in the *c9orf72* knockout neurons (at least three independent experiments, $N > 4$ neurons per genotype per experiment, 17 total neurons were quantified for d and e,*, $p < 0.05$; **, $p < 0.01$; ***, $p < 0.001$). (Data from panel (**e**) is reproduced with permission from Ref. [13])

adding 10% total volume of trypsin of fetal bovine serum (FBS) (GE Healthcare HyClone, SH30071.03), i.e., 500 μl of FBS to 5 ml of trypsin, then centrifuge at 500 × g for 5 min. Discard supernatant, then resuspend tissues in 5 ml of pre-warmed Dulbecco's Modified Eagle Medium: Nutrient Mixture F-12, DMEM/F12 (Thermo Fisher Scientific, 11039-021) mixture; comprising of 10 mM HEPES, pH 7.4 (GE Healthcare HyClone, SH30237.01), 1% penicillin-streptomycin, and 10 U/ml DNase I (Sigma-Aldrich, DN25). Gently triturate tissues 50 times using a long Pasteur pipette. Pass cell mixture through a cell strainer and add 10% total volume of DMEM/F12 of FBS, i.e., 500 μl of FBS to 5 ml of DMEM/F12 mixture. Centrifuge cell mixture at 500 × g for 5 min. Carefully aspirate as much supernatant as possible without disturbing the cell pellet. Resuspend cells in 5 ml of pre-warmed Neurobasal Medium-A (Thermo Fisher Scientific, 21103049) supplemented with 2% B27 supplement (Thermo Fisher Scientific, 17504044), 1% penicillin-streptomycin, GlutaMax (Thermo Fisher Scientific, 35050061), sodium pyruvate (Thermo Fisher Scientific, 11360070), and 50 mM HEPES, pH 7.4. Count cells and dilute to 10^5 cells/ml supplemented with Neurobasal Medium-A. Add 1 ml of cell mixture carefully to each well. Allow cells to settle down and grow in 37 °C incubator, 5% CO_2 for 24 h before checking attachment of cells under the light microscope. Change half of supplemented Neurobasal Medium-A every 3–4 days.

3.2 Visualization and Quantification of Neuronal Morphology and Spine Density

3.2.1 Hippocampal Neuron Transfection Using Lipofectamine 3000

Prepare Lipofectamine 3000 mixture; dilute 1 μl Lipofectamine 3000 in 25 μl Neurobasal Media-A for each well of a 24-well plate. Prepare DNA mix; dilute 500 ng of DNA in 25 μl Neurobasal Media-A and 1 μl of P3000 reagent for each well and incubate mixture at room temperature for 5 min. Meanwhile, replace neuronal culture media with 500 μl of fresh Neurobasal Media-A (neat, without supplements). Add the DNA mix to the Lipofectamine 3000 and gently flick tube to mix, then incubate at room temperature for 10 min. Add 50 μl of DNA/Lipofectamine mixture to each coverslip dropwise. Return plate to 37 °C, 5% CO_2 incubator for 2.5 h (up to 4–5 h for younger and healthier neurons). Transfer coverslips to a new well containing 500 μl of fresh Neurobasal Media-A supplemented with 2% B-27 only. Check transfection efficiency under a fluorescence microscope after 24 h. Fix transfected hippocampal neurons using 4% paraformaldehyde (Electron Microscopy Sciences, 15713) and 4% sucrose (VWR, VWRC0335) for 20 min at room temperature. Rinse coverslips thrice with 1× phosphate-buffered saline, PBS, followed by mounting of coverslips on glass slides using Prolong™ Gold Antifade Mountant with DAPI (Thermo Fisher Scientific, P36931).

3.2.2 Sholl Analysis and Spine Counting

Sholl analysis and spine counting can be done with the Fiji plug-in for ImageJ. Adjust the image (8-bit) threshold prior to analysis. Manually choose the starting and end point for the analyzed area around the cell with the line tool. Open Sholl analysis under analyze; first define the starting and ending radius as well as the radius step size and finally click analyze to generate the list of intersections. For all dendritic spine analyses, select the regions of the apical dendrites after the first branch point, i.e., secondary dendrite. Acquire z-sections at 0.3- μm intervals and stack the images using a maximum intensity projection. Score the dendritic spine density from three randomly chosen areas per neuron; by dividing the number of spines in the chosen area by the length of the dendrite of that area. For in vitro studies in primary hippocampal neurons, all experiments should be repeated at least three to five times.

3.3 Manipulation and Quantification of Neuronal Autophagy

The status of autophagy can be examined using immunofluorescence and immunoblotting against LC3 and fluorescence protein-tagged version of LC3 (Fig. 2) (*see* **Note 2**).

Because we and others have shown that C9orf72 plays a role in autophagy regulation [13, 33–38], we further established that C9orf72-mediated autophagy is required for neuronal morphogenesis and for maintaining dendritic spine density [13]. To address how autophagy may contribute to neuronal morphogenesis and dendritic density, two independent autophagy inhibitors, SBI-0206965 and Spautin-1, which targets ULK1 and Beclin 1, respectively, were used (Figs. 3 and 4). In both cases, autophagy inhibition leads to reduced dendritic arborization (Figs. 3 and 4).

3.3.1 Autophagy Induction and Inhibition with the Use of Small Molecules

Autophagy can be induced in neurons by amino acid and nutrient starvation. Supplemented Neurobasal Media-A is completely removed and replaced with Earle's Balanced Salt Solution, EBSS (Life technologies, 14155-063). Treat cells for 1 and/or 2 h at 37 °C, 5% CO_2; do not treat cells for longer than 2 h (*see* **Note 3**).

Autophagy Induction in Neurons by Nutrient Deprivation

SBI-0206965 (ULK-1 Inhibitor)

Inhibition of autophagy using SBI-0206965 [39] is performed using low concentrations over 7 days as high concentration treatment appears to be toxic to neurons. Prepare a stock solution of 20 mM SBI-0206965 in DMSO, aliquot and store at −20 °C. For treatment, dilute SBI-0206965 to a series of dilutions between 1 and 10 μM in supplemented Neurobasal Media-A. Treat cells for 7 days at 37 °C, 5% CO_2; change half of the culture media with fresh supplemented Neurobasal Media-A containing the respective dilutions of SBI-0206965 on the fourth day of treatment. On the seventh day of treatment, cells can either be lysed using RIPA buffer for protein analysis or transfected using Lipofectamine 3000 for immunofluorescence imaging.

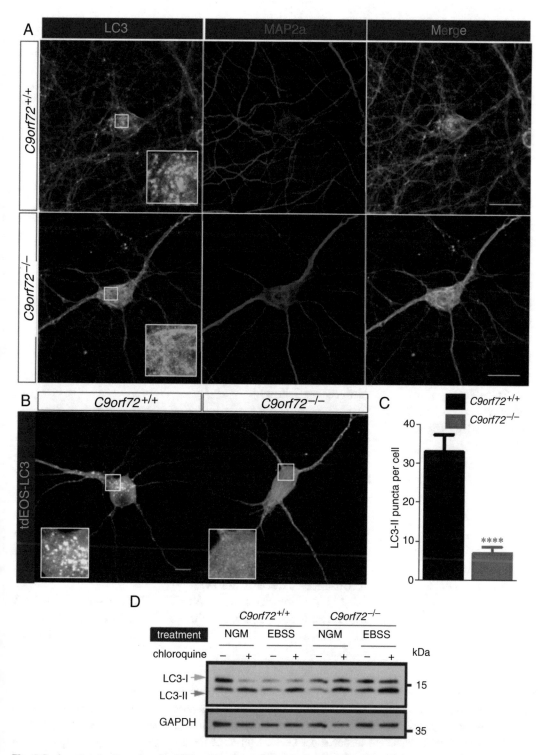

Fig. 2 Reduced autophagy in *c9orf72* knockout neurons. (**a**) Immunofluorescence images of endogenous LC3 and MAP2/MAP2A of primary hippocampal neurons cultured from wild-type and *c9orf72* knockout mice. Scale bar: 20 μm. (**b**) Fluorescent images of wild-type and *c9orf72* knockout hippocampal neurons transfected with

Spautin-1

Inhibition of autophagy is achieved by a combination of Spautin-1 [40] (Sigma, SML0440) and rapamycin. Prepare a stock solution of 10 mM Spautin-1 in DMSO, aliquot and store at −20 °C. For treatment, dilute Spautin-1 and rapamycin in supplemented Neurobasal Media-A at 10 and 0.2 μM, respectively. Treat cells for 4 h at 37 °C, 5% CO_2; after which cells can either be lysed using RIPA buffer for protein analysis or transfected using Lipofectamine 3000 for immunofluorescence imaging.

3.3.2 Genetic Inactivation of Autophagy Machinery

The pharmacological approaches can be further complemented using genetic knockdown and knockout approaches. For example, siRNA against Atg7 can be transfected into primary cortical neurons and achieve effective Atg7 downregulation (Fig. 5) (*see* **Note 4**).

Small Interfering RNA Knockdown in Neurons Using DharmaFECT

Culture cortical neurons in 6-well plate 2 days before siRNA transfection.

Day 0: Cortical Neuronal Culture siRNA Preparation and Transfection (According to Manufacturer's Protocol)

1. Prepare a 5 μM ATG7 siRNA working solution in Neurobasal-A media, as follows:

$$\text{stock} = 100\,\mu M, \quad \text{vol} = 60\,\mu l \qquad \frac{5\,\mu M \times 60\,\mu l}{100\,\mu M} = 3\,\mu l\,\text{in}\,57\,\mu l\,\text{media}$$

$$\text{stock} = 20\,\mu M, \quad \text{vol} = 60\,\mu l \qquad \frac{5\,\mu M \times 60\,\mu l}{20\,\mu M} = 15\,\mu l\,\text{in}\,45\,\mu l\,\text{media}$$

2. In separate tubes, dilute the ATG7 siRNA (Tube 1) and the appropriate amount of DharmaFECT transfection reagent (Dharmacon, T-2001) (Tube 2) with Neurobasal-A media in accordance to Table 1 below:

   *Note: For titration of ATG7 siRNA, a series of siRNA concentrations 0, 5, 10, 25, 50, and 100 μM was used. Different siRNAs have to be titrated accordingly prior to the experiment.

Fig. 2 (continued) photo-convertible GFP (tdEOS)-tagged LC3. Insets show a scaled-up image of the boxed region. Wild-type neurons show visible LC3-II puncta. Scale bar: 50 μm. (**c**) Quantification of LC3-II puncta in wild-type and *C9orf72* knockout hippocampal neurons (at least three independent experiments, $N > 4$ neurons per genotype per experiment, ***, $p < 0.001$, ****, $p < 0.0001$). (**d**) Immunoblots of LC3 and GAPDH on lysates of wild-type (*C9orf72$^{+/+}$*) and c9orf72 knockout (*c9orf72$^{-/-}$*) cortical neurons treated with normal growth medium (NGM) or Earle's balance salt solution in the presence or absence of chloroquine (CQ). LC3-II accumulated in the presence of chloroquine, indicating the autophagy flux is normal in the *c9orf72* knockout neurons. (Data from this figure is reproduced with permission from Ref. [13])

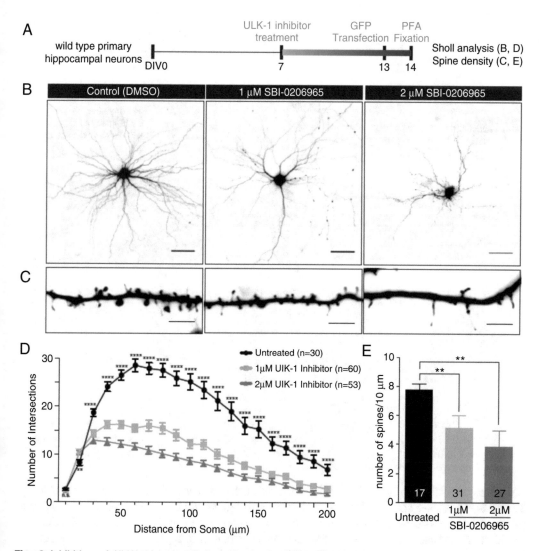

Fig. 3 Inhibition of ULK1 kinase activity reduces dendritic arborization and spine density in wild-type hippocampal neurons. (**a**) Schematic of ULK1 kinase inhibitor (SBI-0206965) treatment in wild-type hippocampal neurons. Wild-type neurons were treated with 1, 2, 5, 10, and 50 μM of SBI-0206965 at DIV7 with a fresh dose every 3 days and transfected with a plasmid encoding GFP at DIV13 to visualize the neuronal morphology. Sholl analysis and spine density was quantified. 10 and 50 μM of SBI-0206965 treatment was toxic to neurons as no neurons survived and thus was excluded from the analysis. (**b**) Images of GFP-transfected hippocampal neurons from control mice treated with DMSO (control), or 1 and 2 μM of SBI-0206965. The images are presented in gray scale and inverted color. Scale bar: 50 μm. (**c**) High-magnification renderings of dendritic segments. The dendritic images were acquired using 0.3-μm step z-sections and then stacked as a maximum projection. Scale bar: 5 μm. (**d, e**) Dose-dependent reduction of dendritic arborization (**d**) and spine density (**e**) in the wild-type neurons treated with SBI-0206965 (at least three independent experiments were performed, total numbers of neurons scored are indicated, *, $p < 0.05$; **, $p < 0.01$; ***, $p < 0.001$, ****, $p < 0.0001$). (Data from this figure is reproduced with permission from Ref. [13])

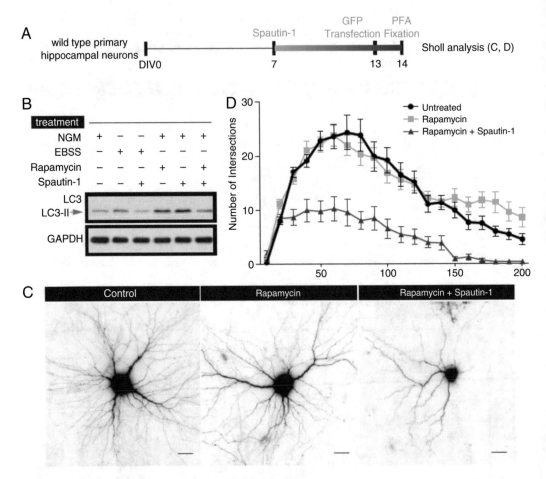

Fig. 4 Autophagy inhibition by Spautin-1 reduces dendritic arborization in wild-type hippocampal neurons. (**a**) Schematic of spautin-1 treatment in wild-type hippocampal neurons. (**b**) Immunoblots of LC3 and GAPDH on lysates of HeLa cells treated with normal growth medium (NGM) or Earle's balance salt solution (EBSS) with or without rapamycin and/or spautin-1. Spautin-1 is more effective in inhibiting LC3-II accumulation when used in combination with rapamycin or starvation treatment. (**c**) Representative images of GFP-transfected hippocampal neurons from control mice treated with DMSO (control), rapamycin (0.2 μM), or spautin-1 (10 μM) and rapamycin (0.2 μM). Wild-type neurons were treated with DMSO, rapamycin, or a rapamycin-spautin-1 combination at DIV13 for 4 h prior to GFP transfection. The images are presented in gray scale and inverted color. Scale bar: 50 μm. (**d**) Sholl analysis of dendritic arborization. N>3 neurons per genotype per experiment from at least three independent experiments were scored. (Data from this figure is reproduced with permission from Ref. [13])

3. Gently mix contents of each tube by pipetting up and down, then incubate for 5 min at room temperature.

4. Combine the siRNA and DharmaFECT solutions, i.e., 200 μl ATG7 siRNA solution +200 μl DharmaFECT solution, and mix by pipetting carefully up and down, followed by incubation at room temperature for 20 min. Add 1600 μl of complete media to the siRNA+DharmaFECT mixture.

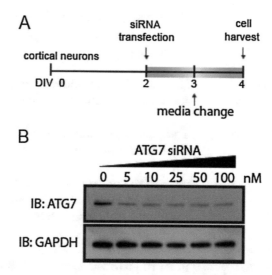

Fig. 5 Knockdown of ATG7 in primary mouse cortical neurons using siRNA. (**a**) Schematic of shRNA treatment of primary cortical neurons. (**b**) Immunoblots of Atg7 under different concentration of Atg7 siRNA

Table 1
Optimization protocol using varying concentrations of siRNA in a 6-well plate

	Tube 1: diluted siRNA (μl/well)		Tube 2: diluted DharmaFECT (μl/well)			
Concentration	Volume of 5 μM siRNA (μl)	NB-A medium (μl)	Volume of DharmaFECT reagent (μl)[a]	NB-A medium (μl)	NB-A with B27 (μl/well)	Total transfection volume (μl/well)
5 nM	2	198	1–10 (5)	199-190	1600	2000
10 nM	4	196	1–10 (5)	199-190	1600	2000
15 nM	6	194	1–10 (5)	199-190	1600	2000
20 nM	8	192	1–10 (5)	199-190	1600	2000
25 nM	10	190	1–10 (5)	199-190	1600	2000
30 nM	12	188	1–10 (5)	199-190	1600	2000
35 nM	14	186	1–10 (5)	199-190	1600	2000
40 nM	16	184	1–10 (5)	199-190	1600	2000
45 nM	18	182	1–10 (5)	199-190	1600	2000
50 nM	20	180	1–10 (5)	199-190	1600	2000
100 nM	40	160	1–10 (5)	199-190	1600	2000

[a]The optimal DharmaFECT reagent amount varies for different cell lines and is affected by the cell density. Easy-to-transfect cells and lower cell densities typically require lower amounts of DharmaFECT reagent

5. Remove culture media from the wells and replace with transfection media. Return plate to 37 °C in 5% CO_2 incubator for 48 h.

Day 4: Check siRNA
Knockdown Efficiency

Harvest cells using radioimmunoprecipitation assay (RIPA) buffer and check for gene knockdown using western blotting.

CRISPR-Cas9-Mediated
Gene Deletion
in Postmitotic Neurons

The ease of genome engineering using CRISPR-Cas9 has revolutionized the biological and biomedical research, including neuroscience [41]. Recent work has shown that CRISPR-Cas9 can also be used to knockout synaptic proteins in primary neurons and slice cultures [42, 43]. The plasmids required for CRISPR-Cas9 are readily available from Addgene (https://www.addgene.org).

Lentivirus Production Using
Lipofectamine 2000

Seed 1.5×10^6 HEK293FT cells in a 10 cm cell culture dish, allow them to grow for 2 days until 80–90% confluent. Prepare Lipofectamine 2000 mixture; dilute 60 µl Lipofectamine 2000 (Thermo Fisher Scientific, 11668030) in 1.5 ml Opti-MEM (Thermo Fisher Scientific, 31985062). Prepare DNA mix; dilute 10 µg of construct vector (e.g., LentiCRISPRv2-mCherry+gRNA or LentiCRISPRv2-mCherry) and 10 µg of each packaging plasmid (FUGW, vSVG and Δ8.9) in 15 ml Opti-MEM. Add the DNA mix to the Lipofectamine 2000 dropwise, gently flick tube to mix, and incubate at room temperature for 20 min. Add transfection mixture to the cells dropwise and return plate to the 37 °C, 5% CO_2 incubator for 4 h. Replace media with transfection mixture to 10 ml of fresh B27 supplemented Neurobasal Media-A. Collect media 48 h post-transfection. Pellet dead cells and debris at 500 g for 10 min. Filter virus-containing supernatant through a 0.45 µM syringe filter, aliquot, and snap freeze in liquid nitrogen.

Lentivirus Infection
in Primary Neuronal Culture

Add virus-containing media (volume needs to be titrated accordingly) to DIV 1- to DIV 2-cultured hippocampal or cortical neurons. Check for gene expression 48 h postinfection. Allow neurons to grow and mature until DIV 14 before harvesting or fixing cells for further analysis.

3.3.3
Immunofluorescence
and Image Acquisition

A Zeiss LSM700 inverted confocal microscope with a 20×/0.8 M27 and a 63×/1.15 N.A. oil immersion objective was used for visualization and image acquisition. Images were captured using an AxioCam MR monochromatic CCD camera (Zeiss) run by Zeiss Zen software. Randomly choose 5–10 EGFP- or mCherry^{CAAX}-labeled neurons from each coverslip with at least three coverslips per genotype. For Sholl analysis, image the entire transfected neuron using the 20×/0.8 M27 objective with z-sections of 0.3- µm intervals. For dendritic spine counting, image 3–5 dendritic segments using the 63×/1.15 N.A. oil immersion objective with z-sections at 0.3- µm intervals. For LC3-II puncta quantification, image the soma of the transfected neuron using the $63\times/1.15$ N. A. oil immersion objective with z-sections at 0.3- μm intervals.

3.3.4 Neuronal Immunofluorescence Staining

Wash the coverslips with $1\times$ PBS with gentle shaking at room temperature. Incubate coverslips in 400 μl of 20 mM glycine (Sigma-Aldrich, 241261) for 5 min at room temperature. Permeabilize cells in 0.3% Triton X-100 in PBS (PBS-T) for 15 min. Block with 5% Donkey Serum (Sigma-Aldrich, S30-100 ml) blocking solution in 0.3% Triton-X 100 (Sigma-Aldrich, X100) in $1\times$ PBS, PBS-T, for 1 h at room temperature. Meanwhile, prepare 400 μl of primary antibody in 1% Donkey Serum blocking solution in $1\times$ PBS for each coverslip. Add 400 μl of diluted primary antibody to each well and incubate at 4 °C overnight. The following day, aspirate the primary antibody and wash coverslips with $1\times$ PBS for 15 min, wash three times with gentle shaking at room temperature. Meanwhile, prepare 400 μl secondary antibody in 1% Donkey Serum blocking solution in $1\times$ PBS for each coverslip. Add 400 μl of diluted secondary antibody to each well and incubate at room temperature for 1–2 h. Protect coverslips from light from this step forth. Aspirate the secondary antibody and wash with $1\times$ PBS once for 10 min with gentle shaking at room temperature. Next, dilute DAPI (Thermo Fisher Scientific, D1306) in $1\times$ PBS at a final concentration of 1 μg/ml. Add 500 μl of DAPI solution to each well and incubate for 10 min at room temperature with gentle shaking. Finally, wash with $1\times$ PBS for 10 min with gentle shaking at room temperature. Check staining efficiency under a fluorescence microscope before mounting on glass slides with Prolong™ Gold Antifade Mountant (Thermo Fisher Scientific, P36930).

3.3.5 Lysate Preparation for Immunoblots

Prepare total protein cell lysate in ice-cold RIPA buffer containing 150 mM NaCl, 1% Triton X-100, 0.5% sodium deoxycholate, 0.1% SDS, and 50 mM Tris (pH 8.0) supplemented with protease and phosphatase inhibitors (Thermo Fisher Scientific, A32959). Lyse cells in buffer for 20 min and mechanically lift cells off the plate surface using a cell scrapper. Collect lysate in a 15 ml tube and centrifuge for 10 min at $17,500 \times g$. Determine protein concentration of the soluble fraction using BCA reagent (Thermo Fisher Scientific, 23225).

3.3.6 LC3-II Puncta Quantification

LC3 is a microtubule-associated protein distributed ubiquitously in a cell. LC3 proteins are able to switch between two forms; LC3-I is the cytosolic, soluble form, whereas LC3-II is the membrane-bound form that is recruited to autophagosomal membranes upon autophagy induction. This method of autophagy tracking is based on the observation that when autophagy is activated, LC3-I is converted (lipidated) to LC3-II, which is subsequently degraded in autolysosomes. Therefore, measuring the levels of LC3-I and LC3-II gives an approximate autophagy status, in which higher

LC3-II levels indicate the presence and abundance of autophagosomes. Inhibiting lysosomal activity during autophagy activation further leads to increased overall LC3-II levels in cells with autophagy activity. The rate of accumulated LC3-II in the above condition (i.e., combining autophagy activation with lysosomal inhibition) provides an estimate on how fast the autophagy process occurs, which is defined as autophagy flux. For quantification of LC3-II puncta, the Fiji plug-in for ImageJ can be employed. Adjust the image (8-bit) threshold to remove the background of diffuse LC3-I prior to analysis. Count the number of LC3-II puncta within the cell body manually. Repeat the experiments at least three to five times.

4   Notes

1. Among the glutamatergic excitatory neurons, granule cells in the dentate gyrus and pyramidal neurons from CA1 and CA3 are the majority neurons in the cultures, which can be distinguished by the expression of distinct transcription factors [44]. To this end, we have performed immunostaining of Prox1 that is specific to dentate granule cells [45], CTIP2 (also known as BLC11B) that selectively labels the granule cells of dentate gyrus and CA1 pyramidal neurons [46], and NeuN as a pan-neuronal marker. Based on the immunoreactivity to Prox1 (DG only), CTIP2 (CA1+DG), and NeuN (CA1 +CA3+DG), we estimated that the dentate granule neurons, CA1, and CA3 pyramidal neurons are present in roughly a 2:1:1 ratio (i.e., 50%:25%:25% of all glutamatergic neurons) in our culture conditions.

2. Due to the elaborated neuronal processes, it is difficult to unequivocally identify LC3 puncta belonging to the same neuron by relying solely on LC3 immunofluorescence. In addition, not only tdEOS-LC3 transfected neurons can be readily identified, but also tdEOS-LC3-positive puncta yield sharper and better signal-to-noise ratio images. Therefore, tdEOS-LC3 transfection is preferred to use for quantification (Fig. 2b, c).

3. Longer treatment (>2 h) causes neuronal death.

4. All genetic manipulation by either siRNA or CRISPR-lentivirus on neuronal cultures should be done on cultures that are not older than DIV 2 for more efficient and robust knockdown.

5 Conclusions

In contrast to the role of autophagy in neurodegeneration [11], how autophagy contributes to neuronal morphogenesis and synaptic biology is less well characterized. Given that autophagosomes are present throughout the axons, dendrites, and soma [47], autophagy is likely to play multiple roles in regulating different aspects of neuronal functions. We hope that the methods described here could serve as a foundation for more exciting studies to come.

Acknowledgments

The authors thank Drs. Sheue-Houy Tyan, Edward Koo, Han-Ming Shen, and Yi-Ping Hsueh for their assistance, critical suggestions, and support for this project. This work was supported by grants to S.-C. L. from the Swee Liew-Wadsworth Endowment fund, National University of Singapore (NUS), National Medical Research Council (NMRC/OFIRG/0001/2016), and Ministry of Education (MOE2016-T2-1-024), Singapore. S.-C. Ling. dedicates this work to the loving memory of Sheue-Houy Tyan.

References

1. Mizushima N, Komatsu M (2011) Autophagy: renovation of cells and tissues. Cell 147:728–741

2. Madeo F, Zimmermann A, Maiuri MC, Kroemer G (2015) Essential role for autophagy in life span extension. J Clin Invest 125:85–93

3. Galluzzi L, Pietrocola F, Levine B, Kroemer G (2014) Metabolic control of autophagy. Cell 159:1263–1276

4. Mizushima N, Yoshimori T, Ohsumi Y (2011) The role of Atg proteins in autophagosome formation. Annu Rev Cell Dev Biol 27:107–132

5. Shen H-M, Mizushima N (2014) At the end of the autophagic road: an emerging understanding of lysosomal functions in autophagy. Trends Biochem Sci 39:61–71

6. Bertram L, Tanzi RE (2005) The genetic epidemiology of neurodegenerative disease. J Clin Invest 115:1449–1457

7. Forman MS, Trojanowski JQ, Lee VM-Y (2004) Neurodegenerative diseases: a decade of discoveries paves the way for therapeutic breakthroughs. Nat Med 10:1055–1063

8. Ling S-C, Polymenidou M, Cleveland DW (2013) Converging mechanisms in ALS and FTD: disrupted RNA and protein homeostasis. Neuron 79:416–438

9. Komatsu M, Waguri S, Chiba T, Murata S, Iwata J, Tanida I et al (2006) Loss of autophagy in the central nervous system causes neurodegeneration in mice. Nature 441:880–884

10. Hara T, Nakamura K, Matsui M, Yamamoto A, Nakahara Y, Suzuki-Migishima R et al (2006) Suppression of basal autophagy in neural cells causes neurodegenerative disease in mice. Nature 441:885–889

11. Menzies FM, Fleming A, Caricasole A, Bento CF, Andrews SP, Ashkenazi A et al (2017) Autophagy and neurodegeneration: pathogenic mechanisms and therapeutic opportunities. Neuron 93:1015–1034

12. Martini-Stoica H, Xu Y, Ballabio A, Zheng H (2006) The autophagy–lysosomal pathway in neurodegeneration: A TFEB perspective. Trends Neurosci 39:221–234

13. Ho WY, Tai YK, Chang J-C, Liang J, Tyan S-H, Chen S et al (2019) The ALS-FTD-linked gene product, C9orf72, regulates neuronal morphogenesis via autophagy. Autophagy 15:827–842

14. Clarke J-P, Mearow K (2016) Autophagy inhibition in endogenous and nutrient-deprived conditions reduces dorsal root ganglia neuron survival and neurite growth in vitro. J Neurosci Res 94:653–670

15. Ban BK, Jun MH, Ryu HH, Jang DJ, Ahmad ST, Lee JA (2013) Autophagy negatively regulates early axon growth in cortical neurons. Mol Cell Biol 33:3907–3919

16. Maday S, Holzbaur ELF (2014) Autophagosome biogenesis in primary neurons follows an ordered and spatially regulated pathway. Dev Cell 30:71–85

17. Stavoe AKH, Hill SE, Hall DH, Colón-Ramos DA (2016) KIF1A/UNC-104 transports ATG-9 to regulate neurodevelopment and autophagy at synapses. Dev Cell 38:171–185

18. Shen W, Ganetzky B (2009) Autophagy promotes synapse development in Drosophila. J Cell Biol 187:71–79

19. Okerlund ND, Schneider K, Leal-Ortiz S, Montenegro-Venegas C, Kim SA, Garner LC et al (2017) Bassoon controls presynaptic autophagy through Atg5. Neuron 93:897–913.e7

20. Soukup S-F, Kuenen S, Vanhauwaert R, Manetsberger J, Hernández-Díaz S, Swerts J et al (2016) A LRRK2-dependent EndophilinA phosphoswitch is critical for macroautophagy at presynaptic terminals. Neuron 92:829–844

21. Shehata M, Matsumura H, Okubo-Suzuki R, Ohkawa N, Inokuchi K (2012) Neuronal stimulation induces autophagy in hippocampal neurons that is involved in AMPA receptor degradation after chemical long-term depression. J Neurosci 32:10413–10422

22. Tang G, Gudsnuk K, Kuo S-H, Cotrina ML, Rosoklija G, Sosunov A et al (2014) Loss of mTOR-dependent macroautophagy causes autistic-like synaptic pruning deficits. Neuron 83:1131–1143

23. DeJesus-Hernandez M, Mackenzie IR, Boeve BF, Boxer AL, Baker M, Rutherford NJ et al (2011) Expanded GGGGCC hexanucleotide repeat in non-coding region of C9ORF72 causes chromosome 9p-linked frontotemporal dementia and amyotrophic lateral sclerosis. Neuron 72:245–256

24. Renton AE, Majounie E, Waite A, Simón-Sánchez J, Rollinson S, Gibbs JR et al (2011) A hexanucleotide repeat expansion in C9ORF72 is the cause of chromosome 9p21-linked ALS-FTD. Neuron 72:257–268

25. Kaech S, Banker G (2006) Culturing hippocampal neurons. Nat Protoc 1:2406–2415

26. Beaudoin GMJ, Lee S-H, Singh D, Yuan Y, Ng Y-G, Reichardt LF et al (2012) Culturing pyramidal neurons from the early postnatal mouse hippocampus and cortex. Nat Protoc 7:1741–1754

27. Craig AM, Banker G (1994) Neuronal polarity. Annu Rev Neurosci 17:267–310

28. Dotti C, Sullivan C, Banker G (1988) The establishment of polarity by hippocampal neurons in culture. J Neurosci 8:1454–1468

29. Ryu J, Liu L, Wong TP, Wu DC, Burette A, Weinberg R et al (2006) A critical role for myosin IIB in dendritic spine morphology and synaptic function. Neuron 49:175–182

30. Chen C-Y, Lin C-W, Chang C-Y, Jiang S-T, Hsueh Y-P (2011) Sarm1, a negative regulator of innate immunity, interacts with syndecan-2 and regulates neuronal morphology. J Cell Biol 193:769–784

31. Klionsky DJ, Abdelmohsen K, Abe A, Abedin MJ, Abeliovich H, Arozena AA et al (2016) Guidelines for the use and interpretation of assays for monitoring autophagy (3rd edition). Autophagy 12:1–222

32. Jiang J, Zhu Q, Gendron TF, Saberi S, McAlonis-Downes M, Seelman A et al (2016) Gain of toxicity from ALS/FTD-linked repeat expansions in C9ORF72 is alleviated by antisense oligonucleotides targeting GGGGCC-containing RNAs. Neuron 90:535–550

33. Sellier C, Campanari M-L, Julie Corbier C, Gaucherot A, Kolb-Cheynel I, Oulad-Abdelghani M et al (2016) Loss of C9ORF72 impairs autophagy and synergizes with polyQ Ataxin-2 to induce motor neuron dysfunction and cell death. EMBO J 35:1276–1297

34. Webster CP, Smith EF, Bauer CS, Moller A, Hautbergue GM, Ferraiuolo L et al (2016) The C9orf72 protein interacts with Rab1a and the ULK1 complex to regulate initiation of autophagy. EMBO J 35:1656–1676

35. Yang M, Liang C, Swaminathan K, Herrlinger S, Lai F, Shiekhattar R et al (2016) A C9ORF72/SMCR8-containing complex regulates ULK1 and plays a dual role in autophagy. Sci Adv 2:e1601167

36. Sullivan PM, Zhou X, Robins AM, Paushter DH, Kim D, Smolka MB et al (2016) The ALS/FTLD associated protein C9orf72 associates with SMCR8 and WDR41 to regulate the autophagy-lysosome pathway. Acta Neuropathol Commun 4(1):51

37. Jung J, Nayak A, Schaeffer V, Starzetz T, Kirsch AK, Müller S et al (2017) Multiplex image-based autophagy RNAi screening identifies SMCR8 as ULK1 kinase activity and gene expression regulator. Elife 6:2

38. Ugolino J, Ji YJ, Conchina K, Chu J, Nirujogi RS, Pandey A et al (2016) Loss of C9orf72 enhances autophagic activity via deregulated mTOR and TFEB signaling. PLoS Genet 12:e1006443

39. Egan DF, Chun MGH, Vamos M, Zou H, Rong J, Miller CJ et al (2015) Small molecule

inhibition of the autophagy kinase ULK1 and identification of ULK1 substrates. Mol Cell 59:285–297

40. Liu J, Xia H, Kim M, Xu L, Li Y, Zhang L et al (2011) Beclin1 controls the levels of p53 by regulating the deubiquitination activity of USP10 and USP13. Cell 147:223–234

41. Heidenreich M, Zhang F (2016) Applications of CRISPR–Cas systems in neuroscience. Nat Rev Neurosci 17:36–44

42. Incontro S, Asensio CS, Edwards RH, Nicoll RA (2014) Efficient, complete deletion of synaptic proteins using CRISPR. Neuron 83:1051–1057

43. Straub C, Granger AJ, Saulnier JL, Sabatini BL (2014) CRISPR/Cas9-mediated gene knockdown in post-mitotic neurons. PLoS One 9: e105584

44. Williams ME, Wilke SA, Daggett A, Davis E, Otto S, Ravi D et al (2011) Cadherin-9 regulates synapse-specific differentiation in the developing hippocampus. Neuron 71:640–655

45. Iwano T, Masuda A, Kiyonari H, Enomoto H, Matsuzaki F (2012) Prox1 postmitotically defines dentate gyrus cells by specifying granule cell identity over CA3 pyramidal cell fate in the hippocampus. Development 139:3051–3062

46. Simon R, Brylka H, Schwegler H, Venkataramanappa S, Andratschke J, Wiegreffe C et al (2012) A dual function of Bcl11b/ Ctip2 in hippocampal neurogenesis. EMBO J 31:2922–2936

47. Kulkarni VV, Maday S (2017) Compartment-specific dynamics and functions of autophagy in neurons. Dev Neurobiol 78:298–310

Chapter 10

Correlative Light and Electron Microscopy (CLEM): Bringing Together the Best of Both Worlds to Study Neuronal Autophagy

Jurgen Kriel, Dumisile Lumkwana, Lydia-Marie Joubert, Martin L. Jones, Christopher J. Peddie, Lucy Collinson, Ben Loos, and Lize Engelbrecht

Abstract

Autophagy is a key protein degradative pathway for primarily long-lived proteins and damaged organelles, contributing to cellular proteostasis. Neurons rely on a particularly effective autophagy machinery, due to their high metabolic activity, ATP demand, and terminally differentiated nature. Autophagy was originally identified with electron microscopy (EM), and it has remained one of the most important tools to study autophagy and its pathway intermediates. However, distinguishing between specific autophagic structures, such as autophagosomes, autolysosomes, and lysosomes, remains challenging, requires substantial expertise, and is often complicated by the large degree of membrane and cargo complexity, underpinned by the heterogeneous electron density within these structures. To accurately identify specific components of the autophagic machinery, correlative light and electron microscopy (CLEM) has emerged as a powerful tool. To discern between autophagic components, specific fluorescent tags can be assigned to macromolecules and structures of interest, and correlated to the ultrastructural detail and subcellular context provided by the electron micrograph. In doing so, a large degree of subjectivity is eliminated, leading to a more accurate characterization and depiction of the total cellular autophagic response and its biological role. This chapter will outline the advantage of CLEM for the study of neuronal autophagy and provide a methodological workflow for both two-dimensional and three-dimensional CLEM approaches.

Keywords Autophagy, Fluorescence microscopy, Correlative light and electron microscopy, Autophagosomes, lysosome

1 Introduction

Autophagy, from the Greek "self-eating," is an important intracellular pathway for the degradation and recycling of cytoplasmic material. The need for accurate identification of autophagic structures, particularly autophagosomes, autolysosomes, and lysosomes, stems from the highly dynamic nature of macroautophagy (hereafter simply referred to as autophagy) and its relation to multiple regulatory processes within the cell [1–4]. Autophagy plays an

Ben Loos and Esther Wong (eds.), *Imaging and Quantifying Neuronal Autophagy*, Neuromethods, vol. 171,
https://doi.org/10.1007/978-1-0716-1589-8_10, © Springer Science+Business Media, LLC, part of Springer Nature 2022

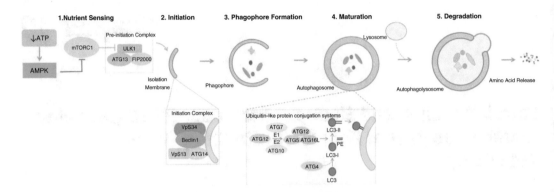

Fig. 1 The autophagy pathway. (1) Decreased amino acid availability and energetic charge (ATP) is sensed by AMP-activated protein kinase (AMPK) and mammalian target of rapamycin (MTOR). Depletion of ATP increases ADP and AMP levels, thereby activating AMPK. AMPK signaling assists in restoring cellular energy levels through autophagy by means of mTOR inhibition and dephosphorylation of the ULK1 activation site. mTROC1 dissociates from the ULK-complex, leading to dephosphorylation of ULK1 and Atg13. Catalytic activation of ULK1 ensues, resulting in ULK-mediated phosphorylation of Atg13 and FIP2000. (2) Activated ULK-complex elicits its kinase activity on key components of the class III PI3K complex (initiation complex). Composed of Vps34, Beclin 1, Vps13, and ATG14, the formation of this multi-domain complex is necessary for the allosteric activation of Vps34. Importantly, Vps34 targets phosphatidylinositol to generate phosphatidylinositol-3-phosphate (PI3P) on the isolation membrane. (3) Elongation is mediated by two ubiquitin-like protein conjugation systems that generate LC3-II. LC3 is lipidated through the actions of a ubiquitin-like protein (UBL) cascade, which involves the E1 (ATG7) and E2 (ATG3) ligases as well as ATG4 protease activity. Formation of the ATG12-ATG5-ATG16L complex facilitates conjugation of LC3-I to PE, producing LC3-II to conclude the elongation reaction. (4) Both LC3-II and the ATG12-ATG5-ATG16L1 complex associate with the elongating membrane, although only LC3-II remains attached to the fully matured autophagosome. (5) Fusion with acidic lysosomes is now possible, allowing for the degradation of cytoplasmic proteins, lipids, and carbohydrates into their respective biosynthetic substrates, which, upon release, may close the regulatory feedback loop

integral role in the clearance of old, toxic, or misfolded proteins, in nutrient sensing, substrate provision, and metabolic integration (Fig. 1) [5, 6]. A basal autophagic activity is inherent and distinct in the mammalian cell, with heightened autophagy contributing fundamentally to cellular stress responses [5, 6]. Dysfunctional autophagy has been implicated in various neurodegenerative pathologies, including Alzheimer's, Parkinson's, and Huntington's disease, drawing major attention to the accurate assessment of neuronal autophagy, its pathway intermediates, and a particular proteinaceous cargo [5]. Many studies, which provide robust evidence for the use of autophagy-inducing drugs to enhance autophagy activity and to clear toxic protein aggregates, rely heavily on microscopy techniques, to confirm whether autophagy was indeed induced in either in vitro, in vivo, or ex vivo models. The accurate assessment and representation of a given autophagic response will hence have a profound impact on its therapeutic validity, particularly in the context of drug discovery studies.

Electron microscopy (EM) and fluorescence microscopy are both critical tools to assess the autophagy pathway. EM, on the one hand, remains one of the most sensitive approaches to detect and also quantify the abundance of autophagy pathway intermediates. It offers high resolution and, importantly, ultrastructural detail and subcellular context. However, it requires major expertise to do so accurately. Moreover, fixation of the sample and a restricted field of view limit the overall information gained from a given biological event, particularly in terms of spatiotemporal and kinetic context. Fluorescence microscopy, on the other hand, provides precise information on specific macromolecules or proteins of interest, and has, through the development of first (GFP-LC3), second (tandem fluorescent constructs, such as mRFP-GFP-LC3), and third (GFP-LC3-RFP-LC3ΔG) generation fluorescent probes, contributed substantially to the assessment of the autophagy pathway and its activity [7]. However, resolution is often limited, and overall molecular context is largely restricted. In both techniques, fluorescence and electron microscopy, inaccurate reporting, such as identifying an autophagosome as an autolysosome or a lysosome as an autolysosome, would greatly misrepresent the true autophagic response. Since the increased presence of autophagosomes may indicate enhanced autophagosome synthesis, and hence autophagy activity, or decreased autophagosome clearance, and hence dysfunction, the accurate identification of autophagy pathway intermediates is essential. Advances in fluorescence microscopy and co-localization analyses have ensured that false-positives remain limited, and as a result, improved quantitative real-time assessment of autophagic flux has been enabled [7, 8]. With regards to conventional transmission electron microscopy (TEM), accurate identification of autophagic structures relies heavily on the expertise of the microscopist and is often circumvented by a less precise use of terminology, such as autophagic vacuoles. Although classical guidelines exist, such as the presence of a double-limiting membrane and the presence of cytoplasmic material, little conformational evidence is provided, unless additional labeling is performed. In this context, correlative light and electron microscopy (CLEM) enables researchers to accurately discern between the main autophagic components through a combination of highly structure-specific fluorescent labeling and high-resolution EM.

In this chapter, we will outline the workflow established in our laboratory to characterize autophagy pathway intermediates in neuronal cells using both 2D and 3D CLEM approaches, with a combination of confocal and super-resolution-structured illumination, and EM.

2   Materials

2.1   Cell Culture

Mouse hypothalamic GT1-7 neuronal cells transfected with a GFP-LC3-RFP-LC3ΔG DNA plasmid (Addgene plasmid ID #84572) and mouse embryonic fibroblast (MEF) cells stably expressing GFP-LC3B protein (the latter a kind gift from Noburu Mizushima, Tokyo University) were utilized. Cells were plated on MatTek gridded coverslip bottom dishes (1×10^4 cells per dish) (MatTek Corporation, Ashland, MA, USA, P35G-1.5-14-CGRD) and cultured in Dulbecco's modified Eagle medium (DMEM) (Life Technologies, 41-965-039) supplemented with 10% fetal bovine serum (Biochrom, S-0615) and penicillin-streptomycin (Life Technologies, 15-140-122) in a humidified incubator at 37 °C with 5% CO_2.

2.2   Microscopy

Confocal and super-resolution structured illumination imaging was conducted using a Zeiss Confocal LSM 780 Elyra PS1 microscope with Zen Black software. EM was conducted using a Zeiss MER-LIN Field Emission Scanning Electron Microscope (FESEM) and a Zeiss Crossbeam 540 Focused Ion Beam SEM (FIB SEM; Zeiss Microscopy, Germany) with Atlas 5 software.

2.3   Sectioning

For sample sectioning, a Leica UC7 ultramicrotome system (Leica Microsystems, Austria) and Ultra 45° 3 mm diamond knife (Diatome US, Hatfield, PA, USA, MS16427) were used to produce ultrathin sections of 70 nm.

2.4   Image Processing Software

To overlay fluorescence and EM micrographs, Fiji (ImageJ) [9] and the EC-CLEM [10] plugin in the Icy image analysis software [11] were used. To visualize structures of interest in 3D FIB-SEM image sets, outlines were annotated using IMOD [12].

3   Methods

3.1   Fluorescence Microscopy: Sample Preparation and Image Acquisition

Prior to confocal imaging, cells were stained with LysoTracker™ Blue DND-99 (Thermo Fisher Scientific, L-7528) (75 nM for 2 h). Thereafter, cells were fixed with 4% paraformaldehyde for 10 min at room temperature. Of note, probes should be chosen that survive fixation and that do not require detergent permeabilization to enter the cell, since detergents strip lipids from membranes and disrupt the ultrastructure required for subsequent EM. After labeling with fluorescent dyes, cells that display the required phenotype must be chosen and their location recorded. To achieve this, a tile scan of a large section of the coverslip was acquired using a Plan-Neofluar 10× objective. When identifying a region of interest (ROI), it is important to record the grid coordinates to facilitate tracking of the region during embedding, and for accurate trimming of the resin block. Once an ROI was identified, z-stacks with a step width of 0.2 μm were acquired of individual cells using a 405 nm diode laser, a 488 nm argon multiline laser (25 mW), and a 561 nm laser line, coupled to a GaAsP detector, utilizing a Alpha Plan-Apochromat 100×/1.46 Oil immersion objective. To enhance the accuracy of subsequent CLEM overlays, it is recommended to also acquire bright-field images, as this assists with the identification of cell outlines more clearly. To ensure optimal overlays, a large z-stack, here 30 image frames, with a maximal step width of 0.2 μm should be acquired to assist in compensating for the difference in z-resolution between fluorescence and electron microscopy. This issue can be further alleviated using super-resolution light microscopy rather than standard confocal fluorescence imaging. In this study, we used super-resolution structured illumination microscopy (SR-SIM) on the Elyra PS1 platform (Carl Zeiss, Germany) to acquire high-resolution z-stacks with 100 nm axial increments. Image frames were collected in five rotations, utilizing an Alpha Plan-Apochromat 100×/1.46 oil DIC M27 ELYRA objective, a 488 nm laser (100 mW), 561 nm laser (100 mW), and an Andor EM-CCD camera (iXon DU 885, Oberkochen, Germany). Images were reconstructed using ZEN software, based on a structured illumination algorithm [13]. For channel alignment, 40 nm fluorescent beads were employed (Zeiss, Tool for Calibration Subresolution, #1783-456). Images of beads were acquired in a z-stacking mode at five phases and three rotations using an Alpha Plan-Apochromat 100×/1.46 Oil DIC M27 objective and 405, 488, and 561 nm laser lines in order to correct for chromatic aberration using the channel alignment algorithm. Subsequently, the images were processed and aligned using ZEN Black Elyra edition software (Carl Zeiss Microscopy).

It should be noted that in the absence of bright-field images, fiducial markers such as fluorescent beads or tracker probes such as mitotracker should be used to provide landmarks that can be identified in both light and electron images to assist with image transformation.

3.2 Electron Microscopy: Sample Preparation

Cells were fixed for 30 min at room temperature with a mixture of 2.5% glutaraldehyde and 4% formaldehyde in 0.1 M phosphate buffer (PB) pH 7.4. Cells were then embedded using the NCMIR protocol, which adds layers of heavy metal to the sample to improve conductivity for block-face electron imaging [14]. Cells were washed in phosphate buffer 5× 3 min on ice and then incubated in 2% reduced osmium (2% OsO_4, 1.5% $K_3Fe(CN)$) for 60 min on ice. Next, cells were washed in dH_2O 5× 3 min at room temperature, to ensure that osmium was sufficiently removed, followed by incubation in 1% TCH for 20 min, and washed in dH_2O 5× 3 min. Once completed, cells were exposed to 2% aqueous OsO_4 in dH_2O for 30 min at room temperature, and washed again in dH_2O for 5×3 min. Cells were thereafter exposed to 1% uranyl acetate overnight at 4 °C. Cells were subsequently washed in dH_2O, 5×3 min, and incubated in lead aspartate for 30 min at 60 °C. Cells were washed in dH_2O, 5×3 min, and lastly, the coverslips carefully detached from the dish using a razor blade and placed in aluminum foil dishes containing dH_2O to avoid drying of cells. Thereafter, cells were dehydrated on ice, using a prechilled alcohol series in the following order: 20%, 50%, 70%, 90%, and $2 \times 100\%$ EtOH for 5 min each step, followed by further incubation of 10 min at room temperature. Once the dehydration was complete, cells were incubated in 50:50 propylene oxide:Durcupan for 60 min. Thereafter, cells were incubated in pure Durcupan for 90 min, twice. Finally, cells were embedded in pure Durcupan in the aluminum dish (for FIB SEM) or by inverting a resin-filled capsule (Agar Scientific Ltd., Essex, UK, G3759) onto the ROI (for 2D SEM). Care should be taken that not too much Durcupan runs out of the capsule into the dish. Dishes were incubated for 48 h at 60 °C in an oven to polymerize the resin.

Once polymerized, the inverted capsule was easily broken away from the coverslip, with the single cell layer embedded in the Durcupan and the pattern of the grid clearly visible with a stereomicroscope. After relocating the grid coordinate containing the ROI, the capsule was trimmed with a razor blade to include only the cells of interest, as previously described [14]. The sample can now be transferred to the FIB SEM for 3D image acquisition or to the ultramicrotome to be sectioned for 2D imaging. For 2D imaging, a Leica UC7 ultramicrotome system (Leica Microsystems, Austria) and Ultra 45° 3 mm diamond knife (Diatome US, Hatfield, PA, USA, MS16427) were used to cut ultrathin sections of 70 nm, collected onto 5 × 5 mm silicon wafer squares (Agar Scientific Ltd., Essex, UK, G3390).

3.3 Electron Microscopy: 2D Image Acquisition/Targeted Array Tomography

After mounting the silicon wafers onto standard 12 mm aluminum SEM specimen stubs using double-sided carbon conductive tape, the cells on the silicon wafers were first observed at a low magnification in the FESEM and compared to the low magnification confocal image (which serves as a map). Depending on the nature and complexity of the sample, various workflows exist for targeted imaging, so as to selectively choose image frames of interest [15, 16]. Here, after confirming that morphological features observed on the confocal image overlap with those of the FESEM, it is advised that cells at the same position on all the subsequent silicon wafers are imaged (see **Note 1**). This adapted targeted array tomography approach, here for selection of localization rather than 3D reconstruction, ensures selectivity for best fitting overlays and decreases the risk of inaccuracy in subsequent processing steps. Ultrastructural features were visualized using a

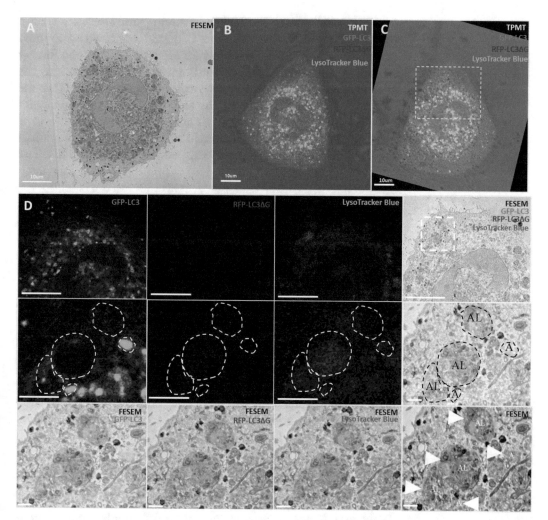

Fig. 2 Outline of workflow for 2D CLEM. (**a**) FESEM micrograph depicting the cell of interest. (**b**) Three-color micrograph and bright-field image channel indicating GFP-LC3-, RFP-LC3-ΔG-, and lysosome-derived signal (blue). (**c**) Transformed micrograph. (**d**) Selective region of interest indicating autophagosomal (A) and autolysosomal (AL) structures, in single and correlated overlaid manner. Scale bar: 10 μm and 1 μm

Zeiss Merlin FESEM operated at 6–7 kV accelerating voltage, 10 nA probe current, and a Zeiss Backscattered Electron Detector (NTS BSD) (Fig. 2a). Electron micrographs were captured in TIFF format at a resolution of 3072 × 2304 pixels, using a pixel-averaging noise-reduction algorithm and SmartSEM software.

3.4 Electron Microscopy: Focused Ion Beam SEM Acquisition

Prior to FIB SEM acquisition, the polymerized samples were removed from the foil dishes by cutting around the coverslip perimeter with a hacksaw, and the glass was removed from the resin surface using liquid nitrogen. After relocation of the ROI using the previously recorded grid coordinates, the blocks were trimmed to a region comprising only a few grid squares, e.g.,

3 × 3, and thinned by further trimming away of excess resin beneath the gridded surface. The trimmed ROIs were mounted on a standard 12.5 mm SEM pin stub using silver paint and further coated with a 10 nm layer of platinum before insertion into the microscope. Care should be taken at this stage to ensure that the grid surface is kept parallel to the stub surface to facilitate easier downstream cross-sectional milling and image acquisition.

The target cells were relocated in the microscope by first imaging the sample surface using the SEM and locating the grid square of interest, and the approximate area within the square. The previously acquired light microscopy data was then used to identify the individual cell of interest. By briefly acquiring an SEM image of the target ROI at a higher accelerating voltage, e.g., 20 kV, it was possible to obtain a map of the cells beneath the surface coating, and by correlating this map to the previously acquired light microscopy data, identify the individual cell of interest and area to be targeted for 3D acquisition.

After preparation for milling and tracking, images were acquired at 5 nm isotropic resolution throughout the ROIs using a 6–10 µs dwell times, dependent on imaging stability and the final size of the selected imaging ROI. The stability of the acquisition is heavily influenced by the environmental conditions close to the microscope, and the overall cycle time from one image to the next. As such, a stable environment and uninterrupted progression are essential for optimal results. During acquisition, the SEM was operated at an accelerating voltage of 1.5 kV with 1 nA current. The EsB detector was used with a grid voltage of 1200 V. Ion beam milling was performed at an accelerating voltage of 30 kV and current of 700 pA.

3.5 Overlay of Fluorescence and Electron Micrographs

To accurately overlay two images, the micrographs must be geometrically transformed to properly align matching features (Fig. 2a–c). This process is more complex for 3D image stacks than for 2D micrographs. Generally, the dataset to be transformed is the fluorescence image stack, since its resolution is lower than that of the EM image stack, and hence loss of data is minimized (*see* **Note 2**). The transformation must take into account the shrinkage and warping that occurs during sample preparation, as well as other artifacts like compression from ultrathin sectioning. Generally, it is advisable to plan which landmarks to use for the alignment ahead of starting the experiment, and for those landmarks to be "off target", i.e., not the structure of interest, to avoid bias in the overlays. In this case, the nuclear probe Hoechst 33342 was used, allowing clear nuclear detail and nuclear edge structures derived from the SIM image to be used as landmarks. Nonbiological landmarks (such as fiducial beads) are not recommended for CLEM of cells and tissues, since membrane permeabilization would be required to get them into the cell, which would disrupt ultrastructure. Fiducial beads

Fig. 3 Outline of workflow for 3D CLEM. (**a**) Maximum projection of super-resolution structured illumination micrograph indicating an autophagosome with corresponding electron micrograph and CLEM overlay (scale bar 1 μm). (**b**) Orthogonal view of FIB SEM stack indicating an autophagosome, with clearly demarcated limiting membrane and cytoplasmic cargo. The resulting model after segmentation of the autophagosome of interest is indicated within the 3D EM dataset **c**. (**c**) Individual planes through the autophagosome from the FIB SEM stack. The 3D SR-SIM render of the autophagosome in (**b**) can be compared with the outline render in IMOD (scale bar 100 nm)

added to the coverslip are unhelpful since they will be lost in the first few resin sections to be cut. Several software solutions exist for this purpose, including BigWarp using Fiji [9, 17], CP2TFORM (MATLAB function), POLYWARP (IDL function), and ec-CLEM [10] (Icy plugin). We have found the ec-CLEM plugin in Icy to be the most user-friendly alternative for both 2D and 3D registration (Figs. 2d and 3a) to produce overlays, which can then be used as a guide for segmentations using IMOD. For detailed instructions on how to implement this specific plugin, the reader is referred to the developer's website (http://icy.bioimageanalysis.org/plugin/ec-CLEM). Importantly, the use of fluorescence markers allowed precise identification of LC3 positive structures, in the absence of requiring lysotrophic signal, and enabled clear distinction between autophagosomes and autolysosomes (Fig. 2d). This in turn allows the assessment of electron density and complexity of cargo material in either structure, revealing autophagosomes characterized by relatively homogeneous, less electron-dense cargo, absent of noticeable lysotrophic material. In sharp contrast, autolysosomes, i.e., GFP-LC3-positive and tracker blue-positive structures are characterized by striking cargo complexity with distinct electron-dense cargo material. Of note, RFP-LC3-Δ positive material may also be present as part of the autolysosome cargo (Fig. 2d), suggesting the effect of temporal, subsequent cytoplasmic sequestration. This is of importance, as it may impact ratiometric analysis for autophagy flux assessment [18]. When performing 3D CLEM, a similar relatively homogeneous electron-dense material can be identified in LC3-positive autophagic structures, with the advantage that the morphological complexity of the autophagosome can be traced and highlighted through the volume, revealing ultrastructural changes in the membrane arrangements and abundance of intra-autophagosomal vesicle-like structures forming part of the cargo (Fig. 3c). This is of importance, due to the cargo-specific turnover in the context of a given autophagosome flux, since now volumetric assessment of specific cargo is enabled [7]. Moreover, through 3D rendering, the autophagosome in its 3D nature may be placed into its ultrastructural context, and hence be visualized in an unprecedented manner (*see* **Note 3**). Additionally, precise volumetric analysis of segmented structures is enabled.

4   Notes

1. *General considerations for 2D CLEM.*

   Although 2D CLEM is considered to be a less complex and technically demanding approach relative to 3D CLEM, it is not without challenges, especially when assessing autophagic structures. Firstly, as the abundance of autophagic structures increases in the fluorescence image stack, the difficulty to accurately identify the corresponding structure in a single plane electron micrograph also rises. Therefore, we advise that initially fluorescent cells with few autophagic structures are selected and that all serial sections are imaged, as this allows better correspondence with the different focal planes. Nevertheless, the possibility of inaccurate signal registration remains, as the resin-embedded cell is sectioned at a slightly different angle than the acquired signal from the laser scanning confocal or SR-SIM microscope. It is, therefore, good practice to acquire multiple cells within a grid space, as this increases the likelihood of finding corresponding electron micrographs at correct focal planes. Moreover, it is important to note that, as described in the workflow here, a SEM may be used for 2D CLEM rather than the more traditional TEM. Although the SEM cannot challenge a TEM for resolution, it allows for many more serial sections to be collected onto a single sample holder, hence favoring an array tomography workflow [19].

2. *Image processing for 2D CLEM.*

   If fiducial markers are not used, it is advised to acquire either a bright-field image or counterstain for organelles or membrane markers, where no permeabilization is required. For super-resolution microscopy without fiducial markers, a membrane or cytoplasmic stain such as CellMask or Phalloidin is recommended. Mitochondria may also be used as landmarks since they are easily labeled using fluorescent dyes, and easily identifiable in EM images by their morphology alone. Nuclear probes are also favorable, best when using SIM-derived data, since nuclear edge and membrane structures are clearly demarcated, which is less the case in confocal-derived micrographs, where fluorescence detail is less resolved. When identifying landmarks, it is easier to transform the fluorescent image by using the bright-field micrograph or detail from a membrane stain to identify morphologically similar features. Once the cell has been transformed to the same orientation as the electron micrograph, the fluorescent labels can be further used for detailed structure identification. Here, overlays were first conducted with a rigid transformation using the ec-CLEM [10] plugin in the Icy image analysis software [11]. Thereafter, ROIs were chosen and processed further through nonrigid transformation using ec-CLEM to enhance overlay accuracy. Selecting a ROI comprising a certain cellular ROI and favorable structural detail, such as prominent autophagic structures, is less time-demanding to overlay compared to accurate analysis of an entire cell. It is also advised that preliminary registration is conducted between the fluorescent stack and the 2D electron micrograph after the initial spatial transformation (Fig. 2). Importantly, even though this step will better outline areas of corresponding focal planes, in our opinion, it is best used as a guide to assist in the final overlay, since the electron micrograph will be scaled to the fluorescent image, which decreases image quality.

3. *Image processing for 3D CLEM.*

FIB SEM image processing:

Although the image stacks resulting from a FIB SEM acquisition are naturally pre-aligned, they benefit significantly from post-processing alignment and noise reduction. Exported TIFF image stacks can be aligned using the "align virtual stack slices" plugin in Fiji, restricted to translational transformations. After alignment, the images can be processed to reduce noise and enhance contrast and sharpness using a combination of a small pixel radius Gaussian blur, followed by 1–2 iterations of using an "unsharp" mask, and a final adjustment of gray levels to enhance contrast if needed. These actions should be tailored to each dataset depending on the inherent image characteristics of each, and extreme care must be taken not to excessively manipulate the image data, thereby removing, obscuring, or adding information, and misrepresenting the output. Processing can be carried out either in Fiji or using commercial solutions such as Amira. The processed image stacks can then be taken forward to 3D registration.

3D CLEM is highly computationally intensive compared to 2D CLEM due to the major increase in file size. It is, therefore, advised that free system memory be at least twice that of the file size to be processed with at least a 6 core CPU. If this is not achievable, compressing the EM-stack and omitting image frames at a constant step-width will make files more manageable, although there will be a noticeable loss in detail. Note that a 3D EM stack from a single cell can comprise of more than 600 slices, and special care should be taken when identifying landmarks. Given adequate computational resources, ec-CLEM performs well for 3D registration. Once a structure of interest has been identified, outlines can be segmented in IMOD [12], which allows for 3D rendering of the final model (Fig. 3). The resulting model file can be used to determine volumetric changes between autophagic structures or to identify sites of membrane fusion zones or damage.

Acknowledgments

This work was supported by the Francis Crick Institute, which receives its core funding from Cancer Research UK (FC001999), the UK Medical Research Council (FC001999), and the Wellcome Trust (FC001999). Moreover, work was supported by the National Research Foundation South Africa (NRF), the South African Medical Research Council (SAMRC), and the Cancer Association South Africa (CANSA).

References

1. Loos B, Toit du A, Hofmeyr J-HS (2014) Defining and measuring autophagosome flux—concept and reality. Autophagy 10:2087–2096. http://www.tandfonline.com/doi/full/10.4161/15548627.2014.973338

2. Guo JY, Teng X, Laddha SV, Ma S, Van Nostrand SC, Yang Y, Khor S, Chan CS, Rabinowitz JD, White E (2016) Autophagy provides metabolic substrates to maintain energy charge and nucleotide pools in Ras-driven lung cancer cells. Genes Dev 30:1704–1717

3. Yin Z, Pascual C, Klionsky D (2016) Autophagy: machinery and regulation. Microb Cell 3:588–596. http://microbialcell.com/researcharticles/autophagy-machinery-and-regulation/

4. Ziviani E, Tao RN, Whitworth AJ (2010) Drosophila parkin requires PINK1 for mitochondrial translocation and ubiquitinates mitofusin. Proc Natl Acad Sci U S A 107:5018–5023.

http://www.pubmedcentral.nih.gov/articlerender.fcgi?artid=2841909&tool=pmcentrez&rendertype=abstract

5. Lumkwana D, du Toit A, Kinnear C, Loos B (2017) Autophagic flux control in neurodegeneration: progress and precision targeting—where do we stand? Prog Neurobiol 153:64–85. https://doi.org/10.1016/j.pneurobio.2017.03.006

6. Yang Z, Klionsky DJ (2010) Mammalian autophagy: core molecular machinery and signaling regulation. Curr Opin Cell Biol 22:124–131

7. Loos B, Klionsky DJ, Du Toit A, Hofmeyr JHS (2020) On the relevance of precision autophagy flux control in vivo–points of departure for clinical translation. Autophagy 16(4):750–762

8. du Toit A, Hofmeyr J-HS, Gniadek TJ, Loos B (2018) Measuring autophagosome flux.

Autophagy 14:1060–1071. https://doi.org/10.1080/15548627.2018.1469590

9. Schindelin J, Arganda-Carreras I, Frise E, Kaynig V, Longair M, Pietzsch T, Preibisch S, Rueden C, Saalfeld S, Schmid B et al (2012) Fiji: an open-source platform for biological-image analysis. Nat Methods 9:676. https://doi.org/10.1038/nmeth.2019

10. Paul-Gilloteaux P, Heiligenstein X, Belle M, Domart M-C, Larijani B, Collinson L, Raposo G, Salamero J (2017) eC-CLEM: flexible multidimensional registration software for correlative microscopies. Nat Methods 14:102. https://doi.org/10.1038/nmeth.4170

11. De Chaumont F, Dallongeville S, Chenouard N, Hervé N, Pop S, Provoost T, Meas-Yedid V, Pankajakshan P, Lecomte T, Le Montagner Y et al (2012) Icy: an open bioimage informatics platform for extended reproducible research. Nat Methods 9:690–696

12. Kremer JR, Mastronarde DN, McIntosh JR (1996) Computer visualization of three-dimensional image data using IMOD. J Struct Biol 116:71–76

13. Heintzmann, R., Cremer CG (1999) Laterally modulated excitation microscopy: improvement of resolution by using a diffraction grating. In: Optical biopsies and microscopic techniques III 3568, pp 185–196

14. Russell MRG, Lerner TR, Burden JJ, Nkwe DO, Pelchen-Matthews A, Domart M-C, Durgan J, Weston A, Jones ML, Peddie CJ, Carzaniga R, Florey O, Marsh M, Gutierrez MG, Collinson LM (2017) 3D correlative light and electron microscopy of cultured cells using serial blockface scanning electron microscopy. J Cell Sci 130(1):278–291

15. Burel A, Lavault MT, Chevalier C, Gnaegi H, Prigent S, Mucciolo A, Dutertre S, Humbel BM, Guillaudeux T, Kolotuev I (2018) A targeted 3D EM and correlative microscopy method using SEM array tomography. Development 145(12):dev160879. https://doi.org/10.1242/dev.160879

16. Wacker I, Spomer W, Hofmann A, Thaler M, Hillmer S, Gengenbach U, Schröder RR (2016) Hierarchical imaging: a new concept for targeted imaging of large volumes from cells to tissues. BMC Cell Biol 17(1):38. https://doi.org/10.1186/s12860-016-0122-8

17. Pietzsch T, Saalfeld S, Preibisch S et al (2015) BigDataViewer: visualization and processing for large image data sets. Nat Methods 12:481–483. https://doi.org/10.1038/nmeth.3392

18. Kaizuka T, Morishita H, Hama Y, Tsukamoto S, Matsui T, Toyota Y, Kodama A, Ishihara T, Mizushima T, Mizushima N (2016) An autophagic flux probe that releases an internal control. Mol Cell 64(4):835–849

19. Micheva KD, Smith SJ (2007) Array tomography: a new tool for imaging the molecular architecture and ultrastructure of neural circuits. Neuron 55(1):25–36

INDEX

Ben Loos and Esther Wong (eds.), *Imaging and Quantifying Neuronal Autophagy*, Neuromethods, vol. 171,
https://doi.org/10.1007/978-1-0716-1589-8, © Springer Science+Business Media, LLC, part of Springer Nature 2022

Printed in the United States
by Baker & Taylor Publisher Services